CBS

DM/MCh Entrance Examinations in Neurology/Neurosurgery

(Also includes Neurosurgery, Neuroradiology & Recent Advances and Some Named Syndromes)

CBS

DM/MCh Entrance Examinations in Neurology/Neurosurgery

(Also includes Neurosurgery, Neuroradiology & Recent Advances and Some Named Syndromes)

Third Edition

2350 MULTIPLE-CHOICE QUESTIONS

By
Dr. M.S. Bhatia
(M.D., MNAMS)
Prof. & Head, Deptt. of Psychiatry,
University College of Medical Sciences,
Dilshad Garden, Delhi - 110 095 (India)

Contributing Editor :
Dr. N. Kaur
(M.D.)
Senior Microbiologist
Dr. R.M.L. Hospital
New Delhi-110 001 (India)

CBS

CBS Publishers & Distributors Pvt. Ltd.

New Delhi • Bengaluru • Chennai • Kochi • Kolkata • Mumbai
Hyderabad • Uttarakhand • Nagpur • Patna • Pune • Jharkhand

ISBN: 978-81-239-2384-0

First Edition: 2006
Second Edition: 2009
Third Edition: 2014
Reprint: 2014, 2015, 2017, 2019

Published by **Satish Kumar Jain** and produced by **Varun Jain** for
CBS Publishers & Distributors Pvt. Ltd.,
4819/XI Prahlad Street, 24 Ansari Road, Daryaganj, New Delhi - 110002
delhi@cbspd.com, cbspubs@airtelmail.in • www.cbspd.com
Ph.: 23289259, 23266861, 23266867 • Fax: 011-23243014

Corporate Office: 204 FIE, Industrial Area, Patparganj, Delhi - 110 092
Ph: 49344934 • Fax: 011-49344935
E-mail: publishing@cbspd.com • publicity@cbspd.com

Branches:
• *Bengaluru:* 2975, 17th Cross, K.R. Road, Bansankari 2nd Stage, Bengaluru - 70 • Ph: +91-80-26771678/79 • Fax: +91-80-26771680
   E-mail: cbsbng@gmail.com, bangalore@cbspd.com
• *Chennai:* No. 7, Subbaraya Street, Shenoy Nagar, Chennai - 600030
   Ph: +91-44-26681266, 26680620 • Fax: +91-44-42032115
   E-mail: chennai@cbspd.com
• *Kochi:* Ashana House, 39/1904, A.M. Thomas Road, Valanjambalam, Ernakulum, Kochi • Ph: +91-484-4059061-65
   Fax: +91-484-4059065 • E-mail: cochin@cbspd.com
• *Kolkata:* 6-B, Ground Floor, Rameshwar Shaw Road, Kolkata - 700014
   Ph: +91-33-22891126/7/8 • E-mail: kolkata@cbspd.com
• *Mumbai:* 83-C, Dr. E. Moses Road, Worli, Mumbai - 400018
   Ph: +91-9833017933, 022-24902340/41 • E-mail: mumbai@cbspd.com

Representatives:

• Hyderabad: 0-9885175004	• Nagpur: 0-9021734563
• Patna: 0-9334159340	• Pune: 0-9623451994
• Jharkhand: 0-9811541605	• Uttarakhand: 0-9716462459

Printed at:
J.S. Offset Printers, Delhi (India)

"Read not to conradict and confute nor to believe and take it for granted, nor to find, talk and discourse but to weigh and consider".

— **Bacon**

ACKNOWLEDGEMENTS

1. Prof. (Dr) Ravi Nehru (MD, DM, DNB), Deptt. of Neurology, GB Pant Hospital, Delhi.

2. Dr. K.S. Anand (MD, DM), Consultant & Head, Deptt. of Neurology, Dr. R.M.L. Hospital, New Delhi.

3. Dr. L.C. Thakur (MD, DM), Director, Prof. & HOD, Deptt. of Neurology, UCMS & GTB Hospital, Delhi.

4. Dr. Deepak Jain (MD, DM, MRCP), Ex. Consultant, AIIMS, New Delhi.

5. Dr. A.K. Bhutani (MS, Mch), Senior Consultant, Deptt. of Neurosurgery, Sir Ganga Ram Hospital, New Delhi.

6. Prof. (Dr) Sunil Kumar (MS), Deptt. of Surgery, UCMS & GTB Hospital, Delhi.

7. Dr. Ravi Gupta (MD), Deptt. of Psychiatry, UCMS & GTB Hospital, Delhi.

8. Dr. Shruti Srivastava (DNB), Deptt. of Psychiatry, UCMS & GTB Hospital, Delhi.

9. Dr. Ramandeep (Mch), Deptt. of Neurosurgery, KGMC, Lucknow.

10. R.C. Jiloha (MD), Director, Prof. & HOD, Deptt. of Psychiatry, GB Pant Hospital, New Delhi.

11. Dr. Savita Jagawat (PhD), Deptt. of Psychiatry, Gwalior Mental Hospital, Gwalior (M.P.).

12. Dr. Tushar Jagawat (DPM, MD), Director, Gwalior Mental Hospital, Gwalior (M.P.).

13. Dr. Sameer (MD, PhD), Deptt. of Radiodiagnosis & Therapy, INMAS, Delhi.

PREFACE

This book is writen wih the aim o outline the major areas of neurosciences (i.e. Neuroanatomy, Neurochemistry, Neurophysiology, Neuropathology, Neuropharmacology, Neuropsychiatry, Clinical Neurology, Neurosurgery and Neuroradiology). This book is not merely an addition to the existing list of MCQ's books but a sincere ambition and an honest attempt to make it a useful and practical companion to both medical undergraduates and postgraduates. It is designed to help a student prepare for course examinations, various postgraduate and services entrance examinations.

The present book consists of ten model test papers (2250 MCQ's) based on he Question Banks of various important examinations (AIIMS, All India and Delhi Entrance Exams, UPSC, AMC, FMGEMS, PLAB, NBER, MRCP, FCGP etc.) We hope that his book will help he candidate in performing better in these examinations.

All suggestions for the modification of this book are welcome and will be duly acknowledged.

— **Authors**

NAMED SIGNS AND SYNDROMES IN PSYCHIATRY AND NEUROLOGY

About this Book

- The new book is a supplement to the textbooks (*"Essentials of Psychiatry"* and *"Short Textbook of Psychiatry"*) and *Dictionary* (of Psychiatry, Psychology & Neurology) by the same author and Publishers.

- Book is divided into two parts. *Part I* contains named signs, syndromes and diseases in Psychiatry and Psychology, *Part II* contains important named Neurological signs and syndromes.

- About *1500* Signs, tests, syndromes, Phenomena and diseases have been listed in alphabetical order.

- Book is useful for students preparing for PG Entrance Exams as well as Postgraduate (MD-Medicine, Psychiatry, Pediatrics) and Postdoctoral (DM/MCH) Exams.

- Useful Appendices include List of Psychological Tests (Appendix I), Cranial Nerves Syndromes (Appendix II) and the Interesting Epileptic Syndromes.

- Edited by: *Dr. M.S. Bhatia* (M.D., Dip. W.P.A.) an experienced teacher and examiner and head, Department of Psychiatry, University College of Medical Sciences and Guru Teg Bahadur Hospital, Dilshad Garden, Delhi-110 095 (India).

Keep your knowledge update with this New book from

CBS Publishers & Distributors

4596/1A, 11, Darya Ganj, New Delhi-110002 (India)

CONTENTS

COMBINED MEDICAL SERVICES EXAMINATION

(By Dr. M.S. Bhatia)

- It contains original solved papers from **1983 onwards**

- All papers are **subjectwise and yearwise** arranged

- Answers are given at the foot

- Detailed **instructions** about this examination are given

- Important **references** to be read are also given

- Each year's paper is classified into two separate parts arranged subjectwise

- Book is suited for *quick* revision before the examination and also to know its pattern

- High chance of repetition in this exam as well as in other UPSC conducted exams

- Read the papers before undertaking preparation (to know the pattern), then read text and again check your knowledge by doing papers

- Also useful for **Civil Services Examination, Part I** as the question bank is same

- Also read book on **Civil Services Examination, Part I (Medical Sciences)** (by same author and publishers) as the **question bank** of two exams may overlap.

- All suggestions are welcome and write to the Editor: **Dr. M.S. Bhatia,** D-1, Naraina Vihar, New Delhi - 110028

Keep your knowledge update with this New book from

CBS Publishers & Distributors
4596/1A, 11, Darya Ganj, New Delhi-110002 (India)

SECTION - I
Important Text in Neurology & Neurosurgery

Dictionary of Psychiatry, Psychology & Neurology

(Dr. M.S. Bhatia)

About this Book

- This Book : A must for :

    - All Medical Undergraduate and Postgraduate Students

    - Various Medical Competitive Examinations

    - Civil Service Examination

- This dictionary is specifically written for easy grasping of Psychiatry, Psychology and Neurology in an easy understandable language.

- About 10,000 words with definitions from the fields of Psychiatry, Psychology and Neurology have been arranged in an alphabetical manner.

- Important terms from Psychiatry and Psychology as related to India have also been included.

- Important appendices include a list of Named Neurological and Psychiatric Syndromes, a list of Psychological Tests and also a list of Mental Hospitals in India with approximately number of beds.

- For the first time, a glossary contains translation of Psychiatric terms from English to Hindi has been made.

Keep your knowledge update with this New book from

CBS Publishers & Distributors

4596/1A, 11, Darya Ganj, New Delhi-110002 (India)

SOME NAMED SYNDROMES/ DISEASES

Addision – Schilder's disease (Adrenoleucodystrophy) : A X-lined recessive disorder, manifest in males and is indistinguishable from Schilder's disease except by the fact that the affected patients show adrenal atrophy.

Aicardi syndrome. A name given to a group of cases, all females, in which severe mental retardation, flexor spasms, chorioretinal lacunae, vertebral and other developmental anomalies associated with corpus collasum agenesis.

Albers – Schonberg disease : A craniotubular modeling disorder, autosomal recessive, characterized by dominant osteopetrosis.

Albright's syndrome (Polyostotic fibrous dysplasia): A syndrome limited to girls, characterized by changes in the long bones, cutaneous pigmentation and precocious puberty and may be accompanied by cranial nerve palsies and/or optic atrophy due to involvement of the bones of the skull base.

Alexander's disease: A degenerative neurological disorder in an infant characterized by mental retardation and progressive megalencephaly. Pathologicallly the condition is characterized by eosinophilic hyaline bodies (Rosenthal fibres) deposited in condition is in subpial and perivascular bands distributed throughout the cerebral white matter.

Alzheimer's disease : The commonest degenerative cause of presenile and senile dementia, characterized by diffuse atrophy of brain most marked in the frontal regions. The onset is gradual and there is early impairment of memory and speech and other neurological signs. It usually develops between the ages of 40 and 60.

Apert's syndrome (Acrocephalosyndactyly) : It is characterized by craniostenosis, oxycephaly, syndactyly, craniofacial dystosis.

Arnoid-Chiari syndrome : A congenital abnormality, which consists of a tongue of cerebellar tissue which projects downwards into the foramen magnum, resulting in obstruction in the cisterna magna at the base of the brain, thus causing hydrocephalus. Spina bifida, meningocele or meningomyelocele are often associated.

Avellis syndrome : A syndrome characterized by ninth and tenth nerve palsy with contralateral hemiparesis and hemianesthesia. It is also known as palatopharyngeal paralysis (in which larynx escapes).

Babinski – Nagotte syndrome : It is characterized by ataxia, Horner's syndrome and contralateral hemiparesis.

Bannwarth's syndrome : It is characterized by multiple mononeuritis with cutaneous erythema, radiculitis, pain and facial nerve involvement.

Bardet-Biedl syndrome : It is characterized by retinitis pigmentosa with features of syndactyly.

Bassen-Kornweig syndrome (abetalipoproteinaemia): This autosomal recessive condition presents in childhood, usually between 2 to 16 years, with atypical retinitis pigmentosa, progressive cerebellar ataxia, mental retardation, peripheral neuropathy and steatorrhoea. The red cells show a typical 'thorny' malformation (acanthocytosis) and the serum level of cholesterol, carotenoids, vitamin A and phospholipids are invariably depressed, with absence of the β lipoprotein moiety.

Becker's disease : The autosomal recessive form of Thomsen's disease.

Behcet's disease: An obscure disorder of ulceration of the mouth and genitalia, iridocyclitis and nervous system involvement with the typical lesions of disseminated encephalomyellin. The common features are headache, progressive dementia, parkinsonian features, ophthalmoplegia and spastic weakness.

Bell's palsy : Facial paralysis of acute onset, due to no known cause and thought to be due to inflammation and swelling of nerve within the stylomastoid foramen. The paralysis is almost always unilateral.

Benedict's syndrome : Paralysis of third cranial nerve on one side with hemianaesthesia, hypertonia, ataxia and tremors on opposite side. Also known as Upper red nucleus syndrome.

Berg Neel syndrome: Macroglobulinaemia with subacute meningitis of lymphocytic type. The other features including strokes, with headache, auditory symptoms and postural hypotension may occur.

Bornholm disease: (Epidemic pleurodynia): An inflammatory disorder of muscle due to Coxackie virus B_5 with pain in the muscles of the trunk and diaphragm giving pain on deep breathing and coughing. It is self liming (a few days).

Bourneville's disease : (Brusfield and Wyatt's disease : Tuberous sclerosis: epiloia): A rare congenital disorder characterized clinically by facial adenoma, mental deficiency and epilepsy, and pathologically by the presence of sclerotic masses within the cerebral cortex and tumours of the brain and viscera.

Brown's syndrome : It is characterized by a mechanical obstruction of the fine movement of the superior oblique tendon so that the clinical picture suggests an isolated palsy of the inferior oblique.

Brown-Sequard syndrome : It comprises pyramidal and posterior column signs in ipsilateral trunk and limbs below the level of lesion, and spinothalamic sensory lesion on opposite side. At the highest level on the side of lesion there is a band of analgesia due to involvement of the root entry zone. Common causes are compression of cord and intramedullary neoplasms, spinal injuries, spinal caries, vascular causes and arachnoiditis.

Canavan's diffuse sclerosis (Van Bogaert – Betrand) : A rare degenerative disorder of early infancy, probably of autosomal recessive inheritance, characterised by progressive mental deterioration, megalencephaly and optic atrophy.

Carpal tunnel syndrome : A cause of pain and paraesthesia involving the flexor aspect of the wrist and those fingers supplied by median nerve. It is characterized by hypaesthesia, Tinel's sign (a tingling sensation radiating into the hand on percussing median nerve at wrist) and Phelan's sign (exaggeration of symptoms after holding the wrist in complete flexion for 30 – 60 seconds).

Charcot's triad : The presence of nystagmus, intention tremor and staccato speech in disseminated sclerosis.

Chediak – Hegashi syndrome: A syndrome characterized by partial albinism, hepatosplenomegaly, lymphadenopathy, mental retardation, cerebellar degeneration with nystagmus and peripheral neuropathy.

Claude's syndrome : Ipsilateral third nerve palsy with crossed cerebellar ataxia, diagnostic of cerebellar ataxia.

Cockayne's syndrome : This rare autosomal recessive disorder of unknown etiology characterized by mental and physical retardation, microcephaly, bird facies, retinal pigmentation, deafness, large hands and feet, a thick skull vault and small pituitary fossa.

Costen's syndrome : Pain on chewing and radiating into the temple or down over the mandible in some cases of temporomandibular arthrosis due to dental malocclusion.

Creutzfeldt's disease : A disorder characterized by combined symptoms referable to the cerebral cortex, basal ganglia and spinal cord. The mental symptoms comprise of confusion, disorientation and impairment of intellect and memory and there are often motor weakness and spasticity, extrapyramidal manifestations and myoclonic jerks. The disease is believed to be a form of transmissible spongiform encephalopathies" caused by an unknown slow virus.

Dandy – Walker syndrome : Atresia of foramina of Luschka and Magendie causing dilatation of 4th ventricle and thus, hydrocephalus.

Dejerine's syndrome : (medial medullary syndrome): Ipsilateral flaccid tongue weakness, contralateral hemiplegia and contralateral loss of position and vibration sense (from infarction of medial lemniscus).

Devic's disease (Neuromyelitis Optica): A disease of unknown aetiology, characterized by an acute transverse myelitis, preceded or accompanied by retrobulbar neuritis, and ending in death or the disease becoming arrested with residual disabilities.

Donohue's syndrome (Leprechaunism): A mental retardation syndrome of unknown cause.

Down's syndrome : The commonest cause of severe mental retardation, occurring in about 1 in 650 live births, due to trisomy (occasionally translocation or mosaicism) of chromosome 21, characterized by mental and physical retardation, a typical mongoloid a appearance of the face with abnormalities of the skull skeleton, hands, feet and palmar creases. Congenital heart lesions and other visceral abnormalities may occur. Those, who survive beyond the age of 40 years, may develop Alzheimer's disease.

Duane's syndrome : A benign congenital abnormality of ocular movement, commoner in females in which there is impaired abduction of the eye.

Eaton-Lambert syndrome : The myasthenic syndrome, usually associated with bronchogenic carcinoma. However it has been reported in patients with rectal, renal, gastric and cutaneous cancer, in leukaemia, reticulum cell carcinoma, malignant thymoma, hypo-and hyperthyroidism, sjogren's disease and sarcoidosis. In some cases no cause is found.

Ehlers – Danlos syndrome : A rate disorder of connective tissue of autosomal dominant inheritance characterized by cutaneous fragility and hyperelasticity, excessive Joint mobility and a bleeding tendency, subarachnoid haemorrhage and multiple aneurysms are common.

Erb type of muscular dystrophy (Limb girdle dystrophy): In this disease, shoulder and pelvic gridles are involved, pseudohypertrophy of gastrocnemii may occur in some patients. It has autosomal recessive inheritance.

Etat Marbre or Status Marmoratus : These terms are used to designate hypermyelination in the basal ganglia. Usually, the aberrant myelin pattern resembles that of veined marble, thus the term etat marbre.

Fahr's disease : An autosomal dominant disease with familial calcification of the basal ganglia, beginning early in life with progressive dementia, convulsions and rigidity.

Fanconi's syndrome : Multiple renal tabular defects with aminoaciduria, hypophosphataemia osteomalacia severe muscular weakness (osteomalacia), myopathy or neurogenic muscular atrophy.

Ferguson-Critchley type ataxia : A hereditary disease characterized by cerebellar ataxia, corticospinal tract and posterior column involvement with ophthalmoplegia.

Field's triad: Cranio-vertebral anomaly characterized by low hair line, short neck and restricted neck movements.

Foster Kennedy syndrome : Common with tumours of the inferior surface of a frontal lobe wihch compresses the adjacent optic nerve causing optic atrophy and as the tumour grows in size, papilloedema occurs in the other eye.

Foux – Chavany – Marie syndrome : The bilateral anterior opernacular syndrome by few pharyngoglosso – masticatory diplegia.

Fovill's syndrome : Facial paralysis, involvement of 6th nerve and crossed hemiplegia (seen in pontine lesions).

Friedreich's ataxia : First signs of disease in adolescence or earlier. Generally bilateral signs and is characterized by dysarthria, nystagmus, loss of tendon jerks, extensor plantars, pes cavus and kyphoscoliosis. Cardiac involvement is also common.

Froin's syndrome : An increased protein content with a normal cell count is described as albumino-cytologic dissociation and is seen in spinal cord obstruction, polyradiculitis (Guillain-Barre syndrome) and chronic meningitis (especially syphilis).

Gaucher's disease : A rare familial disease, chronic in nature, affecting lipoid of kerasin type, characterized by deposition of abnormally large amounts of cerebrosides containing equimolar amount of sphingosine, fatty acid and hexose in the reticuloendothelial cells. The Gaucher and hexose in the reticuloendothelial cells. The Gaucher cell are found in the spleen, liver, lymphnodes and marrow. It is common in infants, children and young adults.

Gerstmann's syndrome : The result of a small, dominant parietal lobe lesion, is the clinical association of finger agnosia, agraphia, left-right disorientation and Inability to calculate.

Gilles de la Tourette's syndrome : It simple vocal as well as multiple motor tics. Moreover, the tics change in distribution and intensity. The vast majority of cases unlike those of habit spasm, are responsive of certain medications but not be traditional psychotherapy or non-specific psychotropic drugs.

Hallervorden-Spatz disease : The rare degenerative involves the globus pallidus and the pars reticulata of the substantia migra. It is apparently hereditary and the clinical features include the onset of the disease at a young age, a motor disorder mainly of the extrapyramidal type, dementia and a relentlessly progressive course leading to death in early adulthood.

Hallgren's syndrome : A syndrome associated with vestibulo-cerebellar ataxia and mental retardation.

Hans – Schuller – Christian syndrome : A chronic or subacute disseminated inflammatory histiocytosis with scattered granulomatous lesions in various tissue containing histiocytes, eosinophils and 'form' cells. There is a characteristic triad of multiple skull defects, diabetes insipidus and exophthalmos, together with liver and spleep enlargements, honeycomb lungs, yellowish brown maculopapular cutaneous lesions, poor growth and sexual development. The aetiology is unknown.

Hartnup disease : Baron, Dent, Harris, Hart and Jepson (1956) first described this hereditary metabolic disorder with renal aminoaciduria with episodic symptoms are of pellagra (photosensitive rash, mental deterioration and cerebellar ataxia")

Heine – Medin disease (Infantile paralysis) : It is a synonym of Poliomyelitis.

Holmes's Adie's syndrome: Unilaterally dilated pupil, reacts promptly to mydriatrics and miotics, but very slowly to light and accommodation, the larger pupil becoming smaller than its fellow. Characteristically occurring in young women, the papillary abnormality is associated with sluggishness or absence of tendon reflexes.

Horner's syndrome : Paralysis of the ocular sympathetic may result from a lesion in the tegmentum of the pons or lateral medullary region. It is associated with miosis, enophthalmos, ptosis and anhidrosis.

Horton's cephalgia (Histamine cephalgia) : Previous name for cluster headache; usually a hemicranial headache, originating from cerebrovascular imbalance, "Clusters refers to the irregular groupings of the headaches. Dozens of brief episodes, one-half to two headaches. Dozens of brief episodes, one – half to two hours long will occur periodically. There is also an ipsilateral (uncommonly bilateral) conjunctival injection with tearing.

Hughlings Jackson syndrome : Characterized by eleventh and twelfth nerve palsies and contralateral hemiplegia.

Huntington's chorea : A hereditary disorder affecting adults and characterized by primary degeneration of the forebrain and caudate nucleus. It often present as involuntary movements, (chorea), mental impairment (dementia) and very rarely the patient develops a terminal Parkinsonian picture replacing the chorea. Chlorpromazine and amphetamine have be found to be successful in controlling chorea.

Hurler's disease (Gargoylism) : A disease of childhood characterized by physical and mental retardation, enlargement of the skull corneal opacities, thickness of the tongue, short stubby fingers and umbilical hernia. There is **hepatosplenomegaly**, Lillac – colloured granules in the leucocytes and reticuloendothelial cells, osteoblastes and histiocytes of the marrow (Railley bodies)

Joseph disease (Machado disease : Azoream disease). An autosomal dominant from of hereditary ataxia with various combinations of cerebellar ataxia, pyramidal and extrapyramidal signs, amyotrophy, dystonia, abnormal eye movements and mild exophthalmos Pathologically, neuronal loss is seen in the substantia nigra, cerebellar dentate nuclei, anterior horns and Clark's column, less often in the striatum.

Kagelberg-Welander syndrome: A heredofamilial form of spinal muscular atrophy simulating muscular dystrophy. Proximal muscles of both the upper and lower limbs are affected first; fasciculations may be seen in some cases.

Kearn – Sayre syndrome : A multisystem disease which has ocular myopathy, retinal pigmentation, cerebellar degeneration and/or cardiomyopathy.

Kernohan – Woltman syndrome or Kernohan's sign : Ipsilateral signs of corticospinal tract dysfunction arise due to pressure of the contralateral crus cerebri against the free edge of the tentorium.

Kline-Levin syndrome : Periodic hypersomnia, is a rare sleep disorder predominantly affecting adolescent boys characterized by excessive sleep for several days to two weeks, three or more times yearly. If patients awake during these prolonged episodes of sleeping, they characteristically eat great quantities of food (bulimia) and then return to sleep. During this period of arousal, patients are slowed in thinking, withdrawn and apathetic.

Klippel Feil syndrome: Basilar impression associated with fusion of the bodies of some cervical vertebra. The patients show mirror movements in the upper extremities.

Kluver-Bucy syndrome: It results due to bilateral anterior lobectomy in animals usually involving the amygdala giving rise to fear, tameness, diminished aggression, a change in dietary habits, sexual overactivity and sometimes, "psychic blindness" but contrary to a widely held belief, no impairment of memory. In humans, it may result due to head injury, Alzheimer's disease, Herpes simplex encephalitis and psychosurgery and memory impairment is invariable.

Kohlmeier-Degos disease (Malignant atrophic papillosis) : A lethal cutaneosystemic vasculopathy, probably not familial, which affects nervous system giving dementia, progressive paresis and sometimes subarachnoid haemorrhage.

Korsakoff's syndrome: It is the mental deficit associated with chronic alcoholism. Its major and frequently exclusive feature is severe memory impairment, especially for relating new information. Contrary to common belief, confabulation is not prominent, frequent or necessary to establish this diagnosis. Classically, the memory deficit is found in the presence of peripheral neuropathy, ataxia and nystagmus (Wernicke-Korsakoff syndrome). Similar memory impairments also are found following Herpes simplex, encephalitis and cerebral anoxia. This is the result of bilateral temporal lobe injury.

Krabbe's disease (Globoid Leukodystrophy) : It results from a deficiency of gactoside – β-galactosidase. The onset is an early infancy. Generalized rigidity, instability and decreased alertness are noted. Infants terminally develop blindness and deafness. It is fatal, usually with in 6 to 12 months of onset.

Kufs' disease: Adult variety of generalized GM, gangliosidosis which does not cause blindness but progressive spastic weakness, epilepsy and/or dementia.

Kugelberg-Welander disease : Hereditary type of motor neuron disease in adolescents, who also have flaccid quardriparesis and atrophic muscles because of anterior horn-cell loss.

Lafora's disease : This degenerative disease begins in childhood or early adulthood and is inherited as an autosomal recessive disorder. The clinical manifestations are convulsive seizures, myoclonic jerks an abnormal EEG and eventually mental deterioration and dementia. The diagnostic morphologic feature is the presence of so called Lafora bodies found throughout the CNS, although they are more common in the basal ganglia and the dentate nucleus, in some cases similar bodies have been seen in the liver and myocardium.

Landouzy – Dejerine dystrophy : (Fascioscapulohumeral dystrophy) : Muscular dystrophy mainly affecting facial and shoulder girdle muscles. The other symptoms like winging and elevation of scapulae, foot drop, may also occur. Progress of illness is relatively slow and it has autosomal dominant inheritance.

Landry-Guillain Barre syndrome (Acute infective polyneuritis) : A syndrome of unknown etiology, characterized by promdromal stage (headache, vomiting, slight fever, pain in back and limbs), latent period (few days to several weeks) and stage of paralysis – motor paralysis (of all four limbs common, dysphagia, ophthalmoplegia and loss of superficial and deep reflexes), sensory symptoms (pain, impairment of all forms of sensations), Cranial nerve paralysis (bilateral VII nerve and also II, III and VI nerves) and rarely retention of urine. Clinical types are mononeuritic, polyneuritic, myelitic (commonest) bulbar, cerebral and recurrent and chronic relapsing types, CSF shows xanthochromia, and albumino-cytologic dissociation.

Laurence – Moon-Biedl syndrome : This disease has mental retardation obesity, hypogonadism and retinitis pigmentosa.

Leigh's disease (Subacute neorotizing encephalomyopathy) : It is inherited as an autosomal recessive trait and is characterized by bilateral symmetric regions of necrosis, cribriform change, vascular proliferation and gliosis in the thalamus, midbrain, brain stem, pons, medulla and spinal cord. The onset, just before the first year of life, manifested by deterioration of motor skills (loss of head control and poor sucking), seizures and myoclonic jerking. Elevated levels of lactic and pyruvate levels and mild PH changes have been reported.

Lennox-Gastaut syndrome : One form of progressive myoclonic epilepsy with 'Petit mal triad' (absences, myoclonic jerking and drop attacks).

Lesch Nyhan syndrome: A rare X-linked disorder confined to males, with mild to moderate mental retardation and choreoathetosis or dystonia which usually begins in the second year of life; this may be followed by the development of spasticity in all four limbs A typical but unexplained feature is self mutilation with biting of the tongue, lips and fingers leading to progressive destruction unless the fingers in particular are protected. Renal calculi, haematuria and renal failure may develop. The serum and urinary uric acid is raised.

Letterer – Siwe disease: The infantile form of Hans – Schulter Christian syndrome.

Louis-Bar syndrome (ataxia telangiectasia): It is characterized by cerebellar ataxia, telangiectasiae in bulbar conjunctiva and skin. It is autosomal recessive.

Lowe's syndrome (cerebrorenal syndrome) : it is characterized by eye changes such as cataract, enophthalmos, corneal opacities, mental retardation and aminoaciduria.

Marchiafava's disease : This is characterized by demyelination of the corpus collasum and optic tracts. The clinical picture is one of acute confusion, hallucinations and excitement. This is often the result of alcohol dependence.

Marie's ataxia : Characterized by cerebellar and pyramidal signs and also a family history.

Marinesco – Sjogren syndrome : A rare autosomal recessive syndrome characterized by somatic and mental retardation, cerebellar ataxia, cataracts and sometimes epilepsy, microcephaly and corticospinal – tract dysfunction.

Maroteaux-Lamy syndrome: An autosomal recessive disease with arylsulphatase B deficiency characterized by progressive spastic paralysis, convulsions, dementia and bulbar paralysis.

Meige's syndrome (Brueghel's syndrome): Blephrospasm – oromandibular dystonia) : A syndrome of unknown aetiology (sometimes precipitated by neuroleptics) which has repetitive sustained spasms or dystonic contractions of facial, mandibular and lingual musculature rarely spreading to cervical laryngeal, pharyngeal and respiratory musculature. Diazepam, antiparkinsonian anticholinergic drugs, lisuride, haloperidol, deanol are some of the drugs found to be effective.

Melkersson's syndrome: Unilateral or bilateral facial palsy with a congenitally fissured tongue.

Meniere's syndrome : characterized by recurrent paroxysms of vertigo, associated with tinnitus and progressive nerve deafness. The cause is unknown but it is associated with dilatation of endolymphatic system.

Menkers' kinky-hair disease (trichopolio-dystrophy): X-linked recessive disorder with early and severe retardation of growth, peculiar white and kinky hair, frequent convulsions and focal cerebral and cerebellar degeneration particularly of gray matter.

Millard – Gubbler syndrome : Nuclear type of facial paralysis with crossed hemiplegia (seen in pontine lesions)

Moebius syndrome: Congenital aplasia of cranial nerve nuclei with bilateral ptosis, absence of elevated of the eyes or lateral rectus paralysis with or without facial paralysis.

Monge's disease (Mountain sickness): It is characterized by fatigue, dyspnoea, clubbing of the fingers, cyanosis, somnolence with polycythemia and haematocrit exceeding 70% pulmonary oedema may occur.

Morvan's syndrome: A hereditary sensory neuropathy with painless perforating ulcers of the feet, with painless fractures or resorption of bones in the feet and even in phalanges.

Moya Moya disease: Takeuchi (1961) and Kudo (1968) first described in Japanese in which the blood vessels supplying the cerebral hemisphere resemble the so called 'rare mirabile" of lower animals. In this there is choreiform syndrome and variable features of cerebral ischemia or subdrural and cerebral haemorrhage.

Niemann Pick disease (Lipoid histic cytosis). A rare familial disease in which sphingomyelin (Ceramide phosphoryl – chalone) and cholesterol accumulate in the reticuloendothelial cell throughout body. There is a retarded physical and mental growth starting early in life and death results from wasting or intercurrent infection within 1 to 3 years.

Parkinsonism : A clinical syndrome comprising impairment of voluntary movements (hypokinesis) rigidity and tremor. It is caused by lesions' in the basal ganglia.

Parry-Romberg syndrome: A disorder of unknown aetiology charactered by progressive wasting of some or all of the tissues of one side of the face.

Pelizaeus-Merzbacher disease : A rare, familial, sexlinked recessive chronic slowly progressive disease occurring in infancy characterized clinically by nystagmus, intention tremors, ataxia, spasticity and visual disturbances.

Pick's disease : A form of degenerative brain disease characterized by atrophy in temporal and frontal regions. The onset is gradual. In early stages, there is loss of normal inhibitions. The more intellectual functions are first to suffer followed by poor memory, disorientation and perseveration.

Pickwickian syndrome: Hypothalamic dysfunction resulting in obesity, day time hypersomnolence (orthodox sleep), nocturnal periodic breathing and hypercapnia.

Ramsay – Hunt syndrome (dyssynergia cerebellaris myoclonica) : It is characterized by progressive cerebellar ataxia with mystagmus, nysarthria, repeated myoclonus, mental retardation (sometime) and progressive dementia.

Rathke's Pouch tumours (Craniopharyngiomas): The tumours arising from the epithelial remnants of the Rathke's Pouch and the common is children. They show very frequent calcification (commonest brain tumour in calcify) and cyst formation. The fluid contains a large amount of cholesterol. The tumours arises in the precede signs of increased tension and may take the form of typical field defects. They may also be polyuria or a true diabetes insipidus.

Refsum's syndrome : Familial disorder of metabolism, Atypical retinitis pigmentosa, cerebellar ataxia and peripheral neuropathy. Thickening of peripheral nerves. Raised CSF protein.

Rett's syndrome : A rare disorder causing developmental stagnation after apparent normal development for about 12 months followed by progressive dementia, autism, truncal ataxia, clumsiness of the hands and acquired microcephaly.

Reye's syndrome: A disease of unknown etiology seen in children (2 months to 15 years of age) associated with vomiting, disturbances of sensorium, seizures, irregular breathing, tachypnoea, hepatomegaly, infrequent jaundice and rapidly developing coma.

Roussy-Levy syndrome : It has hereditary areflexia with distal amyotrophy and ataxia. It shows affinities with Friendreich's ataxia on the one hand and with peroneal muscular dystrophy on the other.

Roussy-Levy syndrome: Friedreich's ataxia associated with peroneal atrophy, familial club foot (pes cavus) and absent reflexes. There may be nystagmus and dysarthria.

Sanger Browns ataxia : Cerebellar signs, optic atrophy, ophthalmoplegia. Same as olivo-ponto cerebellar atrophy.

Schilder's disease : (Diffuse sclerosis or encephalitis periaxialis diffusa): A rapidly progressive condition, characterized pathologically by a widespread demyelination of the cerebral hemispheres and clinically by visual disturbances, convulsions, mental symptoms and sensory disturbances.

Shy-Drager syndrome (Primary atonomic degeneration) : It is usually begins with orthostatic hypotension. Extra pyramidal and cerebellar features occur later associated with impotence, hypotonic urinary bladder and the impairment of perspiration. Atrophy of the iris, the extraocular muscle palsies and peripheral neuropathy may also be seen.

Sjogren – Larsson syndrome : An autosomal recessive disorder characterized by congenital ichthyosis, mental retardation, spasticity, epilepsy, macular degeneration, abnormalities of the teeth and hypertelorism.

Stargardt's disease : A tapetoretinal degeneration or retinal dystrophy, usually of autosomal recessive inheritance and giving rise to progressive loss of central vision usually in individuals between six and 20 years of age.

Steele – Richardson – Obszewski syndrome (Progressive Supranuclear Bulbar Palsy) : It is characterized by having difficulty in voluntary movement of the eyes, usually with impairment of lateral gaze. Later, there is difficulty in executing any vertical eye movement, back rigidity, paroxysmal disequilibrium, expressionless facies and personality changes. Eventually patients become helpless and demented. The onset is usually between the fifth and seventh decades of life. The lesions (neuronal loss and gliosis) are usually in the globus pallidus, subthalamic nuclei, red nucleus, substantia nigra tectum, periaqueductal gray matter and dentate nuclei (the cerebral and cerebellar cortex are usually not involved).

Stotos' syndrome : Mental retardation with cerebral gigantism.

Sturge-Weber disease : It is characterized by haemangioma of cutaneous distribution of the trigeminal nerve in association with one of the leptomeninges.

Sydenham's chorea (St Vitus Dance) : A toxic infective condition, mainly affecting children, that is believed to be due to an infection of the nervous system by acute rheumatism. The onset is often insidious, it is characterized by involuntary movements (sudden jerky quasipurposive) hypotonia, sluggish deep reflexes, excessive excitement, emotional instability and insomnia.

Tangier's disease : A type of alpha lipoproteinemia, orange or yellow discolouration of tonsils, hepatosplenomegaly associated with peripheral neuropathy pigmentosa. This is autosomal recessive.

Taysach's disease (Cerebro macular degeneration) : A heredofamilial disorder characterized clinically by progressive mental failure, blindness and paralysis. Warren Tay (1881) and Bernard Sachs (1887) first described the condition.

Thomsen's disease (Myotonia congenita) : A heredo familial disease characterized by the presence of localized of generalized myotonic symptoms and usually present from birth. The myotonia affects all the skeletal muscles of the body including those of the tongue and mastication. It is autosomal dominant.

Todd's palsy : After recovery from a Jacksonian fit, the parts affected remain paralysed. This is called Todd's palsy and if prolonged for more than an hour or two suggests that there is a structural lesion in or near the cortical representation of the affected part.

Treacher Collins syndrome : It has craniofacial deformity, molar hypoplasia, microgenitalia, deafness and mental retardation.

Usher's syndrome : The combination of congenital deafness and retinitis pigmentosa giving rise to progressive blindness.

Vogt Koyanagi and Harada's disease : It is characterized by uveitis depigmentation of the hair and skin about the eyes, loss of eyelashes and deafness. The CSF findings resemble those of viral meningitis.

von Economo's disease (Postencephalitic Parkinsonism) : This disease is a late sequela of a presumed viral epidemic which spread throughout the world particularly in Eastern Europe, around 1915 and which in some cases induced an acute encephalitis (encephalitis lethargica). In the initial stage, meningeal signs, drowsiness and varying degree of ophthalmoplegia with irregular pupils, poorly reacting to light are seen.

Von Hipple Landau syndrome (Landou's disease): They are true neoplasms of blood vessels and arise from endothelial cells of these vessels. The commonest site is the cerebellum where it usually consists of a cyst around a tumour nodule. This is often associated with cysts in the pancreas and kidneys. This may result in ataxia and papilloedema.

Von Recklinghausen's disease : Multiple neurofibromatosis of the skin.

Wallenberg's syndrome : (Lateral medullary syndrome): Abrupt onset with vertigo, dysphagia, ataxia, ipsilateral anaesthesia of face and contralateral of limbs and trunk, Horner's syndrome, nystagmus and ipsilateral intention tremor. Thrombosis of the posterior inferior cerebellar artery is often the cause.

Weber syndrome : Characterized by a lesion at the upper level of brain stem, resulting in 3rd nerve palsy with crossed hemiplegia.

Wernicke's syndrome : It is an encephalopathy characterized clinically by confusion, nystagmus, extraocular palsies and prostation. It has been shown to be a consequence of thiamine deficiency.

Cranial Nerve Syndromes

Eponym	Site	Cranial Nerves Involved	Common causes
1. Collet – Sicard	Posterior laterocondylar space	IX, X, XI, XII	Tumours of caroid body and secondaries
2. Foix	Sphenoid fissure	III, IV, V, VI	Aneurysms, invasive tumours of sphenoid bone
3. Gradenigo	Apex of petrous bone	V, VI	Petrositis, tumours of petrous bone
4. Jacob	Retrosphenoidal space	II to VI	Larege tumours of middle cranial fossa
5. Tolosa – Hunt	Lateral wall of cavernous sinus	III, IV, V, VI	Aneurysms or thrombosis of cavernous sinus tumors from sinuses and sella turcica
6. Vernet	Jugular foramen	IX, X, XI	Tumours and aneurysm
7. Vilaret Mackenrie Tapia	Posterior parotid space	IX to XII and Bernard – Horner syndrome	Tumours of parotid carotid body, secondaries – lymphnode tumours

CSF in Differential Diagnosis

Disease	Initial Pressure	Appearance	Cell (per Cumm)	Protein (mg%)	Glucose (mg%)	Chlorides (mg%)	Colloidal gold	Comments
Normal	75–200 mm H$_2$O	Clear and Colourless	0–5 lymphocytes	20–45	50–85	700–750	0000110000	Bacteriologically sterile
Bacterial meningitis	Increased	Opalescent to purulent clot	500–20,000 (Polymorphs)	50–1000	0–45	Reduced	Variable	Organisms in sediment or clot, culture positive
Tubercular meningitis	Increased	Opalescent fibrin web, pell web, pell	10–500 (lymphocytes)	45–400	15–75	Reduced	Variable	Sugar and chloride values falling progressively
Viral meningo-encephalitis	Increased	Usually clear	0–200 (mostly lymphocytes)	Normal or Increased	Normal	Normal	Variable	Proved by virus isolation and serological tests
Fungal meningitis	Increased	Clear	Increased (lymphocytes)	Markedly increased	Reduced	Reduced	Variable	Proved by stained smear and culture (usually cryptococci)
Aseptic meningitis (Brain or extradural abscess)	Usually Normal	Clear or turbid often xanthochromic	100–200 (lymphocytes)	Normal or Increased	Normal	Normal	Variable	Culture of CSF negative

Contd...

CSF in Differential Diagnosis (Contd..)

Disease	Initial Pressure	Appearance	Cell (per Cumm)	Protein (mg%)	Glucose (mg%)	Chlorides (mg%)	Colloidal gold	Comments
Traumatic tap (bloody tap)	Normal	Bloody	Many RBCs fresh	Increased	Normal	Normal	Normal	Most blood in first tube, least in last
Syphilis (meningitic acute)	Increased	Clear or turbid; occasional clot	25–2000 (mostly lymphocytes)	45–400	15–75	Normal	1st zone or mid zone curve	CSF and Blood VDRL often positive
Late syphilis	Usually normal	Normal	Normal or Increased	Normal or Increased	Normal	Normal	Depending on activity	Often CSF VDRL positive
Subarachnoid Hemorrhage	Decreased	Bloody, supernatant yellow	Many RBC (crenated or fresh)	Increased	Variable	Normal	Normal	Blood in all CSF specimens
Spinal tumour	Normal or low	Often Xanthochromic	Normal or increased	Normal or Increased	Usually increased	Normal	Variable	Dry tap often

CSF PARAMETERS IN HEALTH AND SOME COMMON DISORDERS

	Normal	Subarachnoid haemorrhage	Pyogenic meningitis	Tuberculous meningitis	Viral meningitis	Multiple sclerosis
Pressure	50-180 mm CSF	Increased	Normal/increased	Normal increased	Normal	Normal
Colour	Crystal Clear	Blood stained xanthochromic	Cloudy	Clear/cloudy	Clear	Clear
Cell count	0-4mm3	Increased red blood cells	Polymorphs 1000-50,000	Lymphocytes 50-5000	Lympho. 10-2000	0-100
Glucose	2/3 blood level	Normal	Decreased	Decreased	Normal	Normal
Protein	<500 mg/l	Increased	Increased	Increased	Normal/increased	Normal/increased
IgG/total protein%	<13%	—	(Not routinely measured)	—	Increased	
IgG index	<0.45	—	(Not routinely measured)	—	Increased	
Oligoclonal IgG bands	Absent	—	(Not routinely measured)	—	Present	
Microbiology	Sterile	Sterile	Organisms on Gram stain and culture	Organisms on ZN stain and culture	sometimes viruses	Sterile

LOCALIZATION OF LESIONS OF THE MOTOR SYSTEM

	Weakness	Tone	Tendon Reflexes	Abnormal Movements	Atrophy
Lower motor neuron (anterior horn cell, root, peripheral nerve)	++	Flaccid	Decreased	No movements	++
Upper motor neuron (cerebral cortex, internal capsule, brain stem, and spinal cord)	++	Spastic	Increased	No movements	± (disuse)
Cerebellum	0	Hypotonia	Decreased	Action Tremor Ataxia	0
Basal Ganglia	0	Rigidity or	Normal	Static Tremor Choreoathetosis	0
Premotor and Frontal Cortex	0	Normal	Normal	Apraxia	0

CHARACTERISTICS OF CSF

Lumbar Puncture, Normal:

I.	*Volume*	–	Total volume (in adults) : 100 – 150 ml
		–	Lateral ventricles : 10 – 15 ml each
		–	Rest of ventricular system : 5 ml
		–	Cranial subarachnoid spaces : 25 ml
		–	Spinal arachnoid spaces : 75 ml
II.	*Production*	–	Normal daily production : 100 ml
		–	Formation : By dialysis, ultrafiltration and secretion by mainly choroids plexus
		–	Absorption : By filtration and osmosis

III. *Pressure*

    A. Reclining
- Newborn : 30 – 80 mm H_2O
- Children : 50 –100 mm H_2O
- Adults : 70 – 200 mm H_2O (AV – 125)

    B. Sitting
- For all ages : Normal if fluid rises to level of foramen magnum (Avg. adult : 350 – 400 mm H_2O)

IV. *Physical characteristics*

Appearance : clear and colourless

Osmolaity : 292 – 297 mosm/ L (=mosm/kg)

Specific gravity: 1.003 – 1.008

pH : 7.31 – 7.34

Sediment : Absent

Cells per Cumm

    Adults : 0 – 5 mononuclears

    Infants : 0 – 20 mononuclears

Differential count : Lymphocytes 60 – 70%, monocytes 30 – 50%, Neutrophils 1 – 3%

V. *Biochemistry*

Alkaline

Total proteins (mostly albumin) : 20 – 45 mg %

Prealbumin : 2 – 6%

Albumin : 56 – 75%

α_1 globulin : 2 – 7%

α_2 globulin : 4 – 12%

β globulin : 8 – 16%

γ globulin : 3 – 12%

IgG : 1.0 – 1.4 mg/dL

IgA : 0.1 – 0.3 mg/dL

IgM : 0.01 – 0.12 mg/dL

Myelin basic protein : < 4 mg/L

Glucose : 50 – 85 mg% (2 – 4 m mol/L)

Chlorides, as sodium chloride :   700 – 740 mg%

$\qquad\qquad\qquad\qquad\qquad$ (120 – 1.0 meq/L)

Colloidal gol (gold sol): 0000110000

Creatinine : 0 – 2.2 mg%

Uric acid (below blood level : 0.3 – 1.5 mg%

Non-protein nitrogen (NPN): 12 – 30 mg%

Amino acids : 30% of blood

Calcium : 4 – 7 mg% (2 – meq/L or 1 – 1.5 m mol/L)

Carbondioxide : 40 – 60 Vol/100 ml (25 meq/L)

PCO_2 : 6 – 7 kPa (45 – 49 mm Hg)

Lactic acid : 8 –27 mg% (0.8 – 2.7 m mol/L)

Magnesium : 3.0 – 3.5 mg % (1 –1.2 m mol/L

$\qquad\qquad$ or 2 – 2.5 meq/L)

Potassium : 12 – 15 mg% (3 – 4 meq/L or 3 – 4 mosm/L)

Phosphate, total : 1.2 – 2.0 mg%

Sodium : 325 mg% (140 meq/L or mosm/L)

Cholesterol : 0.06 – 0.22 mg %

Ammonia : 15 – 47 mol/L 25 – 80 g/dL)

Cisternal Puncture (Other values are as above)

Total proteins : 10 – 25 mg%

Ventricular Puncture (Other values are as above)

Total proteins : 5 –15 mg%

Glucose : 55 – 95 mg%

INTRACRANIAL TUMORS

A) Secondaries (esp, lung)   Commonest
B) Primary

(i)	Glioma	43%
(ii)	Meningioma	18%
(iii)	Pituitary Adenoma	12%
(iv)	Schawanoma	8%
(v)	Chronic pharyngioma	5%
(vi)	Blood vessel Tumor	2%
(vii)	Others	12%

INDICATIONS FOR SKULL X-RAY IN HEAD INJURY

(i) Loss of consciousness/Amnesia
(ii) Focal neurological deficit
(iii) Suspected penentrating injury
(iv) Scalp bruise/Swelling
(v) Alcohol intoxication
(vi) Difficulty in assessing patient

PRIMARY INTRACRANIAL NEOPLASMS*

Neuroepithelial
Astrocytoma, grades I, II, II, and IV
Ependymomas
Oligodendrogliomas
Medulloblastoma
Pinealoma
Papilloma of choroid plexus
Paraphyseal (colloid) cyst
Neurilemoma

Mesodermal
Meningiomas
Hemangioblastoma
Chordoma

Ectodermal
Craniopharyngioma
Pituitary adenomas

Congenital
Epidermoid
Dermoid

* Neoplasms occurring with extreme rarity have been omitted from this classification, those tumors listed in the plural term occur with varying histological types and grades of differentiation.

CAUSES OF ISCHEMIC STROKE

BRAIN STEM LESIONS

Neurologic Sign	Affected Structure
A. MID BRAIN LESIONS	**IPS lateral**
Paresis, eye adduction, elevation	C.N. III
Paresis, depression	C.N. IV
Ptosis,unreative pupil	Descending sympathetic
Paresis upward gaze	Pretectum
Paresis downward gaze	Ventral periaquaductal gray
Ataxia and tremor	Superior cerebellar peduncle
Contralateral	
Ataxia and tremor	Red nucleus
Hemiparesis	Corticobulbar and spinal tracts
Hemihypesthesia	Medial lemniscus, spin othalamic, and quintothalamic tracts
B. *Ipsllateral*	**Affected Structure**
Paresis eye abduction	
Paresis conjugate gaze	Paramedian pontine reticular formation (PPRF)
Internuclear ophthalmoplegia	Medial longitudinal fasciculus
Nystagmus	Vestibular nuclei, PPRF
Facial paresis	C.N. VII
Deafness	C.N. VIII
Ataxia	Middle cerebellar peduncle
Facial hypesthesia	C.N.V.
Paresis chewing	C.N.W.
Horner's syndrome	Descending sympathetic
Contralateral	
Hemiparesis :	Corticcobulbar and spinal tract
Hemilhypestheisa	Medial lemniscus, spinothalamic tract
C. *Medullary lesions*	
Ipsilateral	
Paresis of tongue	C.N. XII
Facial hypalgesia	Spinal trat C.N.V.
Ataxia	Inferior cerebellar peduncle
Nystagmus	Vestibular nuclei
Horner's syndrome	Descending sympathetic

Neurologic Sign	Affected Structure
Loss of taste	Tractus solitarius
Dysphagia, dysarthria	Nucleus ambigus, C.N. IX and X
Contralateral	
Hemiparesis, arm, leg	Corticospinal tract
Hemihypesthesia, arm, leg	Medial lemniscus, spinothalamic tract
Crossed Monoplegia	
Paresis of one	Medial lemniscus, spinothalamic tract
upper extremity and opposite	
lower extremity	Decussaton of pyramids

CHARACTERISTICS OF DYSPHASIAS

Site	Output	Fluency	Para-phasias	Compre-hension	Repetition
Anterior (Broca)	Reduced	Poor	Absent	Retained	Variable
Posterior (Wernicke)	Normal/increased	Good	Present	Impaired	Variable
Arcuate fasciculus (conduction)	Variable	Variable	Mild	Variable	Impaired
Fronto-parietal	Very reduced (global)	Poor	Jargon	Impaired	Impaired

MNEMONICS IN PSYCHIATRY, NEUROLOGY AND ALLIED SCIENCES

(Dr. M.S. Bhatia, M.D., Dip. W.P.A. &
Dr. Tushar Jagawat, D.P.M., M.D.)

About this Book

- This book is must for :

  - All Medical Undergraduates and Postgraduate Students

    - Various Competitive Medical Entrance Examinations

    - Theory, Practical and Viva-Voce Exams

- This book is specifically written to test the knowledge in Psychiatry, Neurology, Psychology and Allied Sciences.

- The book is divided into 4 sections — Section 1 contains Mnemonics in basic Sciences; Section 2 contains Mnemonics in Neurology; Section 3 contains Mnemonics in Psychopharmacology and Section 4 contains Mnemonics in Clinical Psychiatry (including Psychology).

- Each important topic is followed by simple, easy to remember Mnemonic and the details of Mnemonic is given below.

- Topics which are commonly asked in various exams have been selected for this book.

- All contributions are welcome and should be sent to the Editors at : **Dr. M.S. Bhatia**, D-1, Naraina Vihar, New Delhi-110 028.

Keep your knowledge update with this New book from

CBS Publishers & Distributors

4596/1A, 11, Darya Ganj, New Delhi-110002 (India)

SECTION - II
Original Solved
MCQ's

ALL INDIA MD ENTRANCE EXAMINATION
(Conducted by AIIMS)

(By Dr. M.S. Bhatia & Kaur)

- First book with **explanatory** answer to MCQ's of PG Entrance Examination.

- Contains original solved uptodate papers from 1988 onwards.

- Book is **subjectwise** and **year-wise** arranged.

- Each answer is **referenced** to the pages of standard textbook of each subject which also indicates the usefulness of the book for this examination.

- **Explanation to each MCQ** contains very important information of a particular subject around a MCQ.

- **Important Text** is given after each chapter.

- **Important instructions** about this exam are given in the beginning.

- **Important tips** for preparing and useful **references**, examination wise are also indicated in the beginning and also at the start of each chapter.

- **Errata** in question bank is given in the beginning of each chapter.

- Always read this book along with papers of **AIIMS MD Entrance Exam** (given in a separate book by Dr. M.S. Bhatia) because the question banks for these two exams are the same.

- Suited for **quick** revisions after reading a particular subject.

- All suggestions are welcome and write to **Dr. M.S Bhatia,** D-1, Naraina Vihar, New Delhi-110 028.

Keep your knowledge update with this New book from

CBS Publishers & Distributors
4596/1A, 11, Darya Ganj, New Delhi-110002 (India)

ORIGINAL SOLVED MCQ's

1. Rate of formation of cerebrospinal fluid per hour is about:
   A. 5 c.c.
   B. 10 c.c.
   C. 20 c.c
   D. 40 c.c.

2. Function C.S.F., include all except :
   A. Protective
   B. Nutrition
   C. Excretory
   D. Excitatory

3. Apart by the arachnoid villi, CSF is also absorbed around the perineural lymphatics of the following cranial nerves except:
   A. I
   B. II
   C. VI
   D. VII

4. Subarachnoid space ends at the lower border of vertebra:
   A. L
   B. L
   C. S
   D. S

5. Arterial circle of Willis mainly lies in which subarachnoid cistern:
   A. Cisterna magna
   B. Cisterna pontis
   C. Basal cistern
   D. Cisterna ambiens

6. Below what pressure of CSF, its absorption stops:
   A. 125 mm CSF
   B. 96 mm CSF
   C. 65 mm CSF
   D. 42 mm CSF

7. Choroid plexus forms approximately what percentage of total CSF:
   A. 90%
   B. 70%
   C. 50%
   D. 30%

8. Total cholesterol content of CSF is above:
   A. 0.2 mg%
   B. 2.0 mg%
   C. 20 mg%
   D. 100 mg%

9. Which of the following component does not enter an adult brain from plasma:
   A. Magnesium
   B. Phosphate
   C. Urea
   D. Bite

10. If the average value for cerebral blood in an adult is approximately 50 ml/100 g/min then its corresponding value in children in ml/100g/min is:
    A. 40
    B. 50
    C. 75
    D. 100

Ans.: 1. C 2. D 3. C 4. D 5. C 6. C 7. C 8. A 9. D 10. D

11. Because brain tissue and spinal fluid are essentially incompressible, the volume of blood, spinal fluid and brain in the cranium at any time must be relatively constant, This is known as :
    A. Kety principle
    B. Monro-Kellie doctrine
    C. Fick principle
    D. Bell-Magendie Law

12. Which of the following fall during rise of intracranial pressure except:
    A. Pulse rate
    B. Respiration
    C. Cerebral blood flow
    D. Blood pressure

13. The oxygen consumption of oxygen by human brain cerebral metabolic rate for O_2 or (MRO) averages:
    A. 1.0 ml/100g/min
    B. 2.0 ml/100g/min
    C. 3.5 ml/100g/min
    D. 5.0 ml/100g/min

14. The principle that in the spinal cord, the dorsal roots are sensory and the ventral roots are motor is known as:
    A. Bell Magendie law
    B. Fick principle
    C. Kety principle
    D. Monroe-Kellie doctrine

15. Which of the following is a monosynaptic reflex:
    A. Knee jerk
    B. Cremastric reflex
    C. Abdominal reflex
    D. Babinski sign

16. Frquency of alpha rhythm in EEC is increased by all except:
    A. Pyrexia
    B. Glucocoids
    C. Hypoglycemia
    D. PCOs in arteries

17. Desynchronization of EEG pattern is seen in all of the following except:
    A. Opening of eyes
    B. Sensory stimulation
    C. Mental concentration activity
    D. None of the above

18. Sleep spindles characteristically appear in NREM sleep stege:
    A. 1
    B. 2
    C. 3
    D. 4

19. Discharges from nuclei of which part of the brain is believed to intiate REM sleep:
    A. Cerebral cortex
    B. Midbrain
    C. Pons
    D. Medulla

20. Hopping reactions are integrated in which part of the brain:
    A. Cerebral cortex
    B. Midbrain
    C. Medulla
    D. Spinal cord

Ans.: 11. B   12. D   13. C   14. A   15. A   16. C   17. D   18. B
      19. C   20. A

21. **Which of the following organs has both a and b adrenergic receptors:**
    A. Lung
    B. Liver
    C. Intestinal sphincters
    D. Uterus

22. **Which of the following antibiotics is most useful in Pseudomon as infection of meninges :**
    A. Ampicillin
    B. Cloxacillin
    C. Carbenicillin
    D. Chloromycetin

23. **Which of the following drugs may be useful in shortening the course of Guillain Barre syndrome :**
    A. Acyclovin
    B. ACTH
    C. Platelet infusion
    D. Idoxuridine

24. **Prolonged inspiratory cramp followed by expiratory pause in a comatose patient indicates damage to:**
    A. Cerebral cortex
    B. Midbrain
    C. Pons
    D. Medulla

25. **Which of the following is least common in raised intracranial pressure in an infant:**
    A. Suture separation
    B. Papilloedema
    C. Macewen's sign
    D. Crackpot sign

26. **Tonic and clonic convulsions characteristic of major fits are seldom seen in the new born period because of :**
    A. Increased seizure threshold
    B. Poor electrical activity
    C. Musculature is not strong
    D. Statement is not true

27. **Which of the following is the most common cause of pseudotumor cerebri:**
    A. Tetracycline therapy
    B. Vitamin A intoxication
    C. Steroid withdrawal
    D. Idiopathic

28. **The risk of developing epilepsy is high if a child has all of the following except:**
    A. Febrile seizure lasts an house or so and are more than four in a year
    B. Focal in distribution
    C. EEG is abnormal after a few days
    D. None of the above

29. **Convulsions following peretusis is ascribed to:**
    A. Subarachnoid hemorrhage
    B. Anoxia
    C. Bacteremia
    D. Fever

Ans.: 21. D 22. C 23. B 24. C 25. B 26. B 27. D 28. D
       29. B

30. The initial treatment of choice for chronic subdural hematoma in an infant is :

A. Bilateral subdural tap on alternate days if in each tap more than 15 c.c. of fluid

B. Bilateral subdural tap if less 15 c.c. of fluid is removed in each tap

C. Antibiotics and observe

D. Surgical excision

31. Which of the following type of Jack knife convulsion (myoclonus) is most common:

A. Flexor Type   B. Lightning type

C. Extensor spasms   D. Flexpr + Extensor

32. Which of the following is incorrect about rheumatic chorea (sydenbam chorea) :

A. Movements are irregular, nonrepetitive, quasipurposive and involuntary

B. Movements may be limited to one half of the body

C. Movements may affect fingers, hands, extremities and face

D. Phenytoin may be effective for controlling chorea.

33. All of the following are the precipitants of Sydenham chorea except:

A. Attention   B. Sleep

C. Stress   D. Excitement

34. Which of the following aetiologies is believed to be the cause of acute cerebellar ataxia:

A. Head trauma   B. Anoxia

C. Autoimmune   D. Viral

35. Which of the following tract is rarely involved in Friedreich ataxia:

A. Dorsal spinal   B. Pyramidal

C. Spinocerebellar   D. Anterior spinothalamic

36. If a patient with tubercular meningitis receiving Rifampicin, INH and Pyrazinamide develops paresthesia, the treatment of choice is :

A. Steroids   B. BCG

C. Pyridoxine   D. CT scan to rule out abscess

37. Cisternal puncture in comparision to lumbar puncture contains all of the following biochemical abnormalities except:

A. Less protein   B. Less sugar

C. Less pressure   D. Less cells

Ans.: 30. A   31. A   32. D   33. B   34. D   35. D   36. C   37. B

38. **Retrobulbar neuritis is seen in :**
   A. Tabes dorsalis
   B. Syrinobulbia
   C. Multiple sclerosis
   D. Phenytoin toxicity

39. **For 1°F increase in body temperature, cerebral blood flow falls by:**
   A. 1%
   B. 5%
   C. 7%
   D. 10%

40. **Injury to medical cord of brachial plexus results in injury to all except:**
   A. Medial cutaneous nerve of forearn
   B. Medial part of median nerve
   C. Axillary nerve
   D. Ulnar nerve

41. **Bilateral extensors (Babinski sign) are seen in :**
   A. Peroneal muscle dystrophy
   B. Infective polyneuropathy
   C. Brain tumor
   D. Subarachnoid haemorrhage

42. **Which of the following is wrong about meningioma:**
   A. Parasaggital tumors
   B. Reactive hyperostosis seen
   C. Flat
   D. About one fifth of all brain tumors

43. **Not an early sign of extramedullary spinal tumor:**
   A   Bladder involvement
   B   Touch and proprioception lost on the side of lesion
   C   Muscle dystrophy below lesion
   D   None of the above.

44. **Diphenylhydantoin causes all of the following except:**
   A. Ophthalmoplegia
   B. Nystagmus
   C. Megaloblastic anaemia due to B_{12} deficiency.
   D. Lymphoma like syndrome

45. **Not seen in midbrain lesion:**
   A. Crossed hemiplegia
   B. Upward gaze palsy
   C. Dilatation of pupil
   D. Decerebrate rigidity

46. **Automatic bladder is caused by lesion at :**
   A. $S_2 S_2 S_2$
   B. Above S_2
   C. Upper motor neurone
   D. Spinal shock

47. **All are the causes of disseminated CNS lesion except:**
   A. Tuberculosis
   B. Cysticercosis
   C. Malignancy
   D. None of the above

Ans.: 38. C   39. C   40. C   41. C   42. C   43. A   44. C   45. D
      46. C   47. D

48. **Transient syndrome of vertigo, diplopia, slurred speech and paresthesias is seen in:**
    A. Anterior communicating artery aneurysm
    B. Middle cerebral artery thrombosis
    C. Basilar artery insufficiency
    D. Carotid artery stenosis

49. **Demyelinisation within the CNS may be a feature of:**
    A. Post vaccinational   B. Viral diseases
    C. Vitamin deficiencies   D. All of the above.

50. **Hydrocephalus may result from:**
    A. Absence of foramina of Luschka
    B. Aqueductal stenosis
    C. Post meningeal adhesions in the basal cisterns
    D. Degeneration of corpus collasum

51. **In muasthenia grvis, all of the following treatments may be used except:**
    A. Plasmapheresis   B. Surgery
    C. Cholinergic drugs or steroids D. Adrenergic drugs

52. **Tuberous sclerosis is characterised by all of the following except:**
    A. Mental retardation   B. Renal tumors and facial nevi
    C. Epilepsy   D. None of the above

53. **An obese patient has loss of proprioceptive sensation, pains' paresthesias, motor weakness and loss of reflexes. He is most likely suffering from:**
    A. Addison's dieses   B. Myxoedema
    C. Biabetes mellitus   D. Cushing's disease

54. **Occlusion of right posterior cerebral artery is most likely to cause;**
    A. A right sides hemiplegia   B. Total blindness
    C. Homonymous hemianapia   D. VI nerve palsy

55. **A right homonymous hemianopia may be caused by:**
    A. Right optic nerve   B. Right occipital lobe
    C. Right optic radiation   D. Left optic radiations

56. **The aetiology of trigeminal neuralgia is most probably:**
    A. Cold Wind   B. Chewing
    C. Washing of of face   D. Not known

57. **The differential diagnosis of trigeminal neuralgia includes all of the following except:**
    A. Multiple sclerosis
    B. Migraine
    C. Neurpfibroma of V cranial nerve
    D. Bell's palsy

Ans.: 48. C   49. D   50. D   51. D   52. D   53. C   54. C   55. D
      56. D   57. D

58. The treatement of trigeminal neuralgia includes:
   A. Phenytoin   B. Carbamazepine
   C. Clonasepam   D. All of the above

59. In trigeminal neuralgia, the pathological changes includes:
   A. Swelling of nerve   B. Wallerian degeneration
   C. Fibrosis   D. No recognizable change

60. All of the following are true about migraine except:
   A. Common are of onset 10-30 years
   B. More frequent in men
   C. Familial predisposition may have a role
   D. Emotional stress and food stuiffs are common precipitants.

61. The complications of migraine include:
   A. Seizures   B. Psychoilogical disturbances
   C. Subarachnoid haemorrbage   D. All of the above

62. All of the following are true about multiple sclerosis except:
   A. Onset between ages of 20 - 40 years.
   B. Men are affected more than women
   C. Acute infections and trauma are precipitants
   D. Pyelonephritis or cystitis may be the complications

63. A patient with Argyll- Robertson pupil, positive Babinski's sign, increased deep referxes, absent superficial reflexes, retrobullar neuritis has:
   A. Tabes dorsalis   B. Multiple sclerosis
   C. Astrocytoma in cerebellum   D. Pellagra

64. Treatment of choice in syringomyelia is:
   A. Dexamethasone   B. Carbamazepine
   C. Vitamin B_{12} D. Surgery

65. The poisoning with one of the following may cause parkinsonism:
   A. Aresenic   B. Lead
   C. Manganesz   D. All of the above

66. The combinations of disorientation, polyneuritis, loss of recent memory and a tendency to confabulate is most likely due to:
   A. Pernicious anemia   B. Charcot- Marie- Tooth disease
   C. Alcoholism   D. Tabes dorsalis

67. Which of the following is least common symptom in multiple sclerosis:
   A. Leg weekness   B. Visual loss
   C. Incoordination   D. Vertigo

Ans.: 58. D   59. D   60. B   61. D   62. B   63. B   64. D   65. C
      66. C   67. D

68. Muscular wasting can occur in all of the following except:
   A. Syringomyelia   B. Myasthenia gravia
   C. Cushing's syndrome   D. None of the above

69 Involvement of the optic chiasma with visual field defects can occur in all of the following except:
   A. Pituitary adenoma
   B. Craniopharyngioma
   C. Internal carotid artery aneurysm
   D. Falx meningioma

70. In Wilson's disease there is:
   A. An increase in copper excretion in urine
   B. A reductionm of serum ceruloplasmin
   C. Renal involvement
   D. All of the above.

71. The most effective treatment in Parkinsonism includes:
   A. Levodopa   B. Benserazide
   C. Brnzhexol   D. Combination of the above.

72. Which of the following is most useful in the diagnosis of stroke:
   A. CT scan   B. EEG
   C. Cerebral angiography   D. Clinical examination

73. Which of the following is most important in the treatment of head injury:
   A. Maintenance of airway   B. I/V fluid
   C. Antibiotics   D. Vasopressors

74. Neuropathy may be produced by all of the following drug except:
   A. Chloroquine   B. Pyrazinamide
   C. Nitrofurantoin   D. Emetine

75. Sciatica is most commonly produced by:
   A. Ankolysing spondylitis   B. Fracture of lumbar vertebra
   C. Spinal tumour   D. Herniation of intervertebral disc

76. In which of the following muscular dystrophies the life expectancy is not altered:
   A. Duchenne type   B. Erb's type
   C. Landuzy Dejerine type   D. None of the above

77. In the diagnosis of myasthenia gravis, which of the following short acting anticholinesterase is used:
   A. Edrophonium   B. Neostigmine
   C. Pyridostigmine   D. Destigmine

Ans.: 68. D   69. D   70. D   71. D   72. D   73. A   74. B   75. D
      76. C   77. A

78. **A 28 year old labourer develops bilateral foot drop, progressing over one week to paralysis of both legs and trunk. There are no constitutional symptoms. CSF shows albumino leucocytie dissociation. The diagnosis is:**
    A. Lead toxicity
    B. Cyanide poisoning
    C. Guillain Barre sysdrome
    D. Alcoholic neuropathy

79. **Injury to ulnar nerve results in:**
    A. Wrist drop
    B. Inability to appose the thumb
    C. Impaired adduction and abduction of the fingers
    D. Atrophy of muscles of the thenar eminence

80. **The severity and frequency of herpes zoster increases:**
    A. With increase in age
    B. In patients with leukaemia
    C. In patients with lymphoma
    D. All of the above.

81. **The pathological changes are common in multiple sclerosis in all of the following brain areas except:**
    A. Around the ventricles
    B. Spinal cord
    C. Gray matter of cerebrum
    D. Optic nerves

82. **The commonest presentation of multiple sclerosis is :**
    A. Weakness of one or more limbs
    B. Unilateral retrobulbar neuritis
    C. Ataxia
    D. Epilepsy

83. **In most instances in multiple sclerosis, patient recovers within:**
    A. 1 month
    B. 1-3 months
    C. 3-6 months
    D. 6-9 months

84. **A tingling or electric shock like" sensation which radiates into the arms, down in the back or into the legs when the patient flexes the head is known as:**
    A. Lasegue's sign
    B. Fabere sign
    C. Romberg's sign
    D. Barber chair sign

85. **The most common cause of Barber chair sign is:**
    A. Spinal cord compression
    B. Vitamin B_{12} deficiency
    C. Multiple sclerosis
    D. Parkinsonism

86. **The commonest emotional reation in multiple sclerosis is:**
    A. Depression
    B. Euphoria
    C. Dementia
    D. Delirium

Ans.: 78. C   79. C   80. D   81. C   82. A   83. B   84. D   85. C
      86. B

87. The diagnosis of multiple sclerosis depends on:
    A. Clinical demonstration of lesion occurring at different times.
    B. Evoked potential
    C. Electrophoresis of CSF protein
    D. All of the above.

88. All of the following are good prognostic factors in multiple sclerosis except:
    A. When retrobulbar neuritis is the initial presentation.
    B. Middle aged spinal form
    C. Female sex
    D. Late age of onset.

89. Which of the following promotes the more rapid and complete recovery during acute exacerbation of multiple sclerosis.
    A. Dexamethasone   B. Baclofen
    C. Azathioprine   D. Spinal cord stimulation

90. Which of the following is the characteristic pathological change in Paralysis agitans:
    A. Lewy bodies
    B. Patchy cortical atrophy
    C. Atrophy of the globus pallidus
    D. Depigmentation of the susbstantia nigra

91. The most common symptom for which a patient with Parkinsonism seeks medical advice:
    A. Akinesia   B. Tremors at rest
    C. Rigidity   D. Blephrospasm

92. Physiological tremor differs from parkinsonian tremor by all of the following except:
    A. Much faster
    B. Tends to occur only in one plane
    C. Lacks abduction/adduction movement of the thumb
    D. None of the above

93. Levodopa is most effective in relieving:
    A. Tremors   B. Rigidity
    C. Hypokinesis   D. All of the above

94. Which of the following drug is contraindicated along with levodopa:
    A. Imipramine   B. Amantadine
    C. Tranylcypromine   D. Benztropine

Ans.: 87. D   88. D   89. A   90. A   91. B   92. D   93. C   94. C

95. **Following are the symptoms of on/off phenomenon seen during levodopa therapy except:**
    A. Hypotonia
    B. Unsteadiness of gait
    C. Akinesis
    D. Tremors

96. **Stereotactic thalamotomy is indicated in parkinsonism if there is following symptom unresponsive to drugs:**
    A. Akinesia
    B. Unlateral tremor
    C. Rigidity
    D. Hypotonia

97. **Commonest type of Parkinson's disease is:**
    A. Paralysis agitans
    B. Drug induced
    C. Encephalitis
    D. Head injury

98. **The following neurological diseases are inherited as autosomal recessive except:**
    A. Ataxia telangiectasia
    B. Friedreich's ataxia
    C. Wilson's disease
    D. Fascioscapulo-humeral muscular dystrophy

99. **In Wilson's disease, the lesions are most characteristic in:**
    A. Substantia nigra
    B. Corpus striatum
    C. Caudate nucleus
    D. Putamen

100. **Kayser – Fleischer ring, characteristic of Wilson's disease, has following colours except:**
    A. Golden brown
    B. Yellow
    C. Orange
    D. Green

101. **Moro's reflex is abnormal after:**
    A. 4 weeks
    B. 8 weeks
    C. 12 weeks
    D. 20 weeks

102. **Tram-line calcification is seen in:**
    A. Ependymoma
    B. Thrombosed cerebral vein
    C. Meningioma
    D. Sturge-Weber syndrome

103. **Paradoxical sleep consists of :**
    A. REM, high spike, slow waves
    B. REM, sharp wave, fast rhythm
    C. NREM, delta waves
    D. NREM, high spike, theta waves

104. **In spinal shock, reflexes are:**

	Automatic	*Somatic*
A.	Unaffected	Decreased
B.	Unaffected	Increased
C.	Increased	Increased
D.	Decreased	Decreased

Ans.: 95. D   96. B   97. A   98. D   99. B   100. C   101. C   102. D
103. A   104. A

105. **The commonest calcified brain mass in a child in suprasellar region is:**
    A. Craniopharyangioma   B. Medulloblastoma
    C. Tuberculoma   D. Meningioma

106. **The commonest intramedullary spinal tumor is:**
    A. Chordoma   B. Meningioma
    C. Astrocytoma   D. Epennymoma

107. **Wrist drop is seen in palsy of:**
    A. Ulnar nerve   B. Radial nerve
    C. A + B   D. Median nerve

108. **Best treatment for established kernicterus is:**
    A. Steroids   B. Antibiotics
    C. Exchange transfusion   D. No treatment

109. **Drug used to control St. Vitus dance (sydenham's Chore) is:**
    A. Penicillin   B. Haloperidol
    C. Imipramine   D. Phenytoin

110. **The movements of Huntington's chorea are more prominent in the:**
    A. Face   B. Trunk
    C. Lower extremities   D. Upper extremities

111. **Following drungs may produce tics except:**
    A. Amphetamine   B. Levodopa
    C. Neuroleptics   D. Propranolol

112. **The commonest type of motor neurone disease:**
    A. Progressive bulbar palsy
    B. Amyotrophic lateral scloerosis
    C. Progressive muscular atrophy
    D. All are equally common.

113. **The earliest manifestation of amyotrophic lateral sclerosis is:**
    A. Wasting of small muscles of hands
    B. Spasticity of the legs
    C. Extensor plantar
    D. Brisk Jaw Jerk

114. **A 55 year old man presents with signs of upper motor neurone lesion in lower limps and signs of lower motor neurone lesion in upper limps CSF, Wasserman reaction is negative. He has:**
    A. Tabes dorsalis
    B. Progressive muscular atrophy
    C. Amyotrophic lateral sclerosis
    D. Subacute combined degeneration

Ans.: 105. A   106. D   107. B   108. D   109. B   110. A   111. D
      112. B   113. A   114. C

115. **All of the following are true about motor neurone disease except :**
    A. The commonest age of onset is in the sixth decade' rare before 40 years
    B. It is twice as frequent in males
    C. The initial manifestations are commonly unilateral
    D. Fasciculations are pathognomic of the condition

116. **The cause of death in motor neurone disease is related to :**
    A. Respiratory system
    B. Cardiovascular system
    C. Digestive system
    D. Urogenital system

117. **The neural defects of syringomyelia most commonly affects:**
    A. Upper cervical region
    B. Lower cervical region
    C. Upper thoracic region
    D. Lower thoracic region

118. **The most characteristic feature of syringomyelia is:**
    A. Charcot's joints
    B. Excruciating pain in the arms
    C. Dissociated sensory loss
    D. Wasting of small muscles of hands

119. **All of the following are the manifestation of syringobulbia except:**
    A. Kyphoscoliosis
    B. Horner's syndrome
    C. Pyramidal tract signs in the legs
    D. None of the above

120. **The tumors which are associated with neurofibromatosis include all except:**
    A. Gliomas
    B. Ependymoma
    C. Meningiomas
    D. Phaeochromocytoma

121. **Which of the following is not seen in pellagra:**
    A. Dysarthria
    B. Dysphagia
    C. Diarrhoea
    D. None of the above

122. **The presentations of central pontine myelinolysis in alcoholics include all of the following except:**
    A. Tetraparesis
    B. Dysphasia
    C. Anarthria
    D. Ophthalmoplegia

123. **The most common presentation of subacute combined degeneration is:**
    A. Loss of position sense
    B. Loss of vibration sense
    C. Paraesthesiae of the toes
    D. Ataxia.

Ans.: 115. D 116. A 117. B 118. C 119. D 120. B 121. D
122. D 123. C

124. The best test for diagnosing subacute combined degeneration completely recovers with therapy:
A. Peripheral neuropathy
B. Acid output
C. Clinical examination
D. Serum B_{12} levels

125. Which of the following symptom of subacute combined degeneration completely recovers with therapy:
A. Peripheral neuropathy
B. Ataxia
C. Spasticity
D. Dementia

126. Which of the following is implicated in the causation of tropical spinal ataxia:
A. Lead
B. Manganese
C. Cyanide
D. Arsenic

127. The treatment of burning feet syndrome includes vitamin: .
A. B_1
B. B_2
C. B_{12}
D. B-complex

128. The most common initial symptom of cord compression is :
A. Pain
B. Paraesthesiae
C. Loss of deep touch
D. Loss of proprioception

129. Lumbar puncture in suspected cord compression should:
A. Always be done because it is diagnostic
B. Always be done because it may be therapeutic
C. Never be done unless facilities for neurosurgery present
D. Always be done above level of compression.

130. The differential diagnosis of cord compression includes all of the following except:
A. Multiple sclerosis
B. Syringomyelia
C. Motor neurone disease
D. None of the above

131. All of the following factors may influence the prognosis of cord compression except:
A. Severity
B. Duration
C. Cause
D. Type pf sensory loss

132. Disc herniation is most common between vertebrae:
A. $C_5 - C_6$
B. $C_6 - C_7$
C. $C_8 - T_1$
D. $L_1 - L_2$

133. The cervical disc herniation is most commonly situated:
A. Anteriorly
B. Posteriorly
C. Laterally
D. Any of the above

Ans.: 124. D 125. A 126. C 127. D 128. A 129. C 130. D
131. D 132. B 133. C

134. The commonest presentation of dorsomedial herniation of discs is:
A. Sensory loss in the upper and lower limbs.
B. Sensory loss in the upper limbs and spasticity in the legs
C. Spasticity in the upper limbs and sensory loss in lower limbs.
D. Spasticity in both upper and lower limbs.

135. The acute syndrome of cervical intervertebral disc protrusion is best treated by:
A. Active neck and shoulder exercises
B. Analgesics
C. Intermittent neck traction followed by immobilization
D. Surgery

136. Paraesthesiae and numbness in the Lumbago-Sciatica syndrome is seen in the distribution of nerve root.
A. L_4
B. L_5
C. S_1
D. S_2

137. All are the causes pf carpal tunnel syndrome except:
A. Rheumatoid arthritis
B. Acromegaly
C. Pregnancy
D. Hyperthyroidism

138. The commonest cause of mononeuritis multiplex in India is:
A. Leprosy
B. Rheumatoid arthritis
C. Polyarteritis nodosa
D. Sarcoidosis

139. Which of the following is an uncommon symptom in neuralgic amyotrophy (localized radiculopathy):
A. Severe pain
B. Paralysis of muscles
C. Loss of tendon jerks in the affected limbs
D. Sensory loss.

140. The following drugs may cause neuropathy except:
A. Vincristine
B. INH
C. Steroids
D. Chloroquine

141. Which of the following types of diabetic neuropathy is also known as "diabetic pseudotabes":
A. Distal sensori motor polyneuropathy
B. Diabetic amyotrophy
C. Sensory neuropathy
D. Autonomic neuropathy

142. The following are the common symptom of acute intermittent porphyria except:
A. Confusion
B. Tachycardia
C. Hypotension
D. Colicky Abdominal pain

Ans.: 134. B   135. C   136. C   137. D   138. A   139. D   140. C
141. C   142. C

143. **The following drugs are contraindicated in acute intermittent porphyria except:**
    A. Oral contraceptives
    B. Diazepam.
    C. Alcohol
    D. Aulfonamides

144. **The HLA which is believed to be associated with myasthenia gravis is:**
    A. A_2
    B. B_2
    C. B_{27}
    D. DR-4

145. **The first symptom of myasthenia gravis is:**
    A. Diplopia
    B. Dysphagia
    C. Dysarthria
    D. Dyspnoea

146. **The most frequent symptom of diabetic autonomic neuropathy is:**
    A. Diarrhoea
    B. Impaired cardiovascular reflexes
    C. Impotence
    D. Disturbance in micturition

147. **Drug which sometimes may improve the muscle power in carcinomatous myasthenia is:**
    A. Diazepam
    B. Destigmine
    C. Guanidine
    D. Durabolin

148. **All of the following myopathies are autosomal dominant except:**
    A. Myotonia atrophica
    B. Myotonia congenital (Thomsen's disease)
    C. Limb-girdle type (Scapulohumeral type of Erb)
    D. Facio-scapulo humeral type (Landouzy Dejerine)

149. **Drug most usful in the treatment of hydatid cyst is:**
    A. Thiabendazole
    B. Flumebendazole
    C. Mebendazole
    D. None of the above

150. **All of the following drugs are used in the treatment of myotonia except:**
    A. Procainamide
    B. Quinine sulphate
    C. Diphenylhydantoin
    D. Alcohol

151. **Cerebrospinal fluid replaces which component of body into the CNS:**
    A. Blood
    B. Serum
    C. Plasma
    D. Lymph

152. **Cerebrospinal fluid is mainly formed in the :**
    A. Choroid plexus of 3rd ventricle
    B. Choroid plexus of 4th ventricle
    C. Choroid plexus of lateral ventricles
    D. Capillaries on the surface of the brain and spinal cord

---

Ans.: 143. B   144. B   145. A   146. C   147. C   148. C   149. B
     150. D   151. D   152. C

153. Interventricular formen (between lateral ventricles) is also known as :
A. Foramen of Magendie   B. Foramen of Luschka
C. Foramen of Monro   D. Cerebral aqueduct.

154. Total quantity of CSF in the body is about:
A. 50 c.c   B. 100 c.c
C. 150 c.c   D. 250 c.c

155. "Tela choroidea a sheath surrounding the choroidal plexus is formed by:
A. Dura mater   B. Arachnoid mater
C. Pia mater   D. All of the above

156. CSF in contrast to plasma contains a more of the following substances except:
A. Potassium   B. Magnesium
C. Chloride   D. None of the above

157. Which of the following component has ratio between CSF and plasma as 1.0:
A. Magnesium   B. Chloride
C. PH   D. Osmolality

158. The brain weighs about 1400 g in air but in its "water bath" of CSF, it has a net weight of:
A. 20 g   B. 50 g
C. 100 g   D. 500 g

159. Which of the following component crosses blood brain barrier at the slowest pace:
A. Water   B. Glucose
C. CO_2 D. O_2

160. Which of the following component has the lowest CSF plasma ratio:
A. Uric acid   B. Calcium
C. Protein   D. Glucose

161. All of the following factors influence the total cerebral blood flow except:
A. Arterial and venous pressures at brain leve
B. The viscosity of Blood
C. Sleep
D. Sternuous mental activity

162. Total blood flow in brain ml per minute is approximately:
A. 1500   B. 1000
C. 750   D. 600

Ans.: 153. C   154. C   155. C   156. A   157. D   158. B   159. B
160. C   161. D   162. C

163. **Brain receives what percentage of total cardiac output:**
    A. 8%   B. 14%
    C. 16%   D. 23%
164. **In humans, the approximate reaction time for a knee jerk is:**
    A. 2-5 m sec   B. 5-15 m sec
    C. 20-25 m sec   D. 25-60 m sec
165. **In two point discrimination test, 2 points are perceived as separate on the back if distance is atleast (approximately):**
    A. 3 mm   B. 30 mm
    C. 60 mm   D. 2 cm
166. **Which of the following is known as "Alpha Block" :**
    A. EEG pattern during opening of eyes
    B. EEG pattern during closing of eyes
    C. EEG during sleep
    D. EEG during hyperpyrexia
167. **The frequency of "Sleep spindles " resemble those of:**
    A. Delta waves   B. Theta waves
    C. Alpha waves   D. Beta waves
168. **Children have more of what stage of NREM sleep:**
    A. 1   B. 2
    C. 3   D. None of the above
169. **Which of the following reflex is integrated in mid-brain:**
    A. Tonic neck reflex   B. Tonic labyrinthine reflex
    C. Labyrinthine righting reflex D. Optical righting reflex
170. **Match the following :**

Lesion in	Movement
I) Caudte nucleus	(i) Hemiballism
II) Subthalaic nulei	(ii) Chorea
III) Lenticular nucleus	(iii) Althetosis

    A. I (i)   II (ii)   III (iii)
    B. I (ii)   II (iii)   III (i)
    C. I (iii)   II (i)   III (ii)
    D. I (ii)   II (i)   III (iii)
171. **Which of the following organs is not innervated with beta receptors:**
    A. Skin   B. Heart
    C. Juxtaglomerular cells   D. Liver
172. **Which of the cranial nerve is most notably involved in Guillain Barre syndrome:**
    A. I   B. II
    C. III   D. VII

Ans.: 163. B   164. C   165. C   166. A   167. C   168. C   169. C
        170. D   171. A   172. D

173. In severe cases of Guillain Barre syndrome with associated respiratory paralysis, treatment of choice is:
A. ACTH
B. Acyclovir
C. Plasmapheresis
D. None of the above

174. Oculovestibular response (i.e. if the external auditory canal is irrigated with cold water, the eyes normally deviate towards the stimulated side) is lost in all of the following except:
A. Pontine lesions
B. Labyrinthitis
C. Phenytoin toxicity
D. Amphetamine toxicity

175. Which of the following cranial nerve palsy is common in raised intracranial pressure displacing the brain stem downwards:
A. I
B. III
C. VI
D. VII

176. Pseudotumor cerebri (Benign intracranial hypertension) may be seen in all of the following except:
A. Galactosemia
B. Hypoparathyroidism
C. Thrombosis of superior sagittal sinus
D. Thrombosis of lateral sinus

177. All of the following are the causes of convulsions in a new born on the first day of life except:
A. Birth asphyxia
B. Narcotic withdrawal
C. Tetany
D. Inborn errors of metabolism like phenylketonuria

178. Which of the following is not true about febrile seizures:
A. The seizure does not last for more than 10 minutes
B. Convulsions may be generalized or focal
C. There is no residual weakness
D. They are unusual before the age of 6 months and after 5 years

179. Chronic subdural hematoma (common in age group of 2 to 6 months) is most often due to:
A. Rupture of middle cerebral aretery
B. Rupture of venous sinus
C. Rupture of cortical veins
D. Subarachnoid hemorrhage

180. Which of the following drugs may be used in the treatment of infantile spasms(massive myoclonic seizures or Salaam fits)':
A. Diazepam
B. Phenobarbitone
C. ACTH
D. All of the above

Ans.: 173. C   174. D   175. C   176. C   177. C   178. B   179. C
180. D

181. **In a grand mal epilepsy, muscular rigidity during tonic phase is most marked in which group of muscles:**
    A. Flexors of arms and lower extremities
    B. Flexors of arms and extensors of lower extremities
    C. Extensors of arms and flexors of lower extremities
    D. Extensor of arms and lower extremities

182. **Which of the following is incorrect about sydenham chorea:**
    A. Nearly one third of these patients develop rheumatic vavular lesions
    B. Common in girls and has age of onset 5-15 years
    C. Rheumatic polyarthritis may associated
    D. ASLO titre may not be elevated

183. **The following are the differential diagnosis of Sydenham's chorea except:**
    A. Huntington's chorea   B. Wilson's disease
    C. Hyperthyroldism   D. None of the above

184. **Which of the following virus has been implicated in the causation of acute cerebellar ataxia:**
    A Measles   B Mumps
    C Influenza   E. Epstein Barr

185. **Todd's hemiparesis usually recovers within :**
    A. 1 – 2 days   B. 3 days – 1 week
    C. 1 – 2 weeks   D. 2 – 4 weeks

186. **If a patient of tubercular meningitis develops headache and vomiting, the cause may be determined with:**
    A. Lumbar puncture
    B. Electroencephalography
    C. Blood levels of drugs
    D. CT scan

187. **Squatting test(when the patient is unable to stand from the squatting positing while keeping his hands on head) is useful in the diagnosis of:**
    A. Neurosyphilis
    B. Neurocysticercosis
    C. Subacute combined degeneration
    D. Wernicke's encephalopathy

188. **A patient with head injury and hypotension, most likely has:**
    A. Brain stem Injury   B. Pontin haemorrhage
    C. Subdural haemorrhage   D. Bleeding from other site

Ans.: 181. B   182. C   183. D   184. D   185. C   186. D   187. D
       188. D

189. Autonomous bladder is produced by injury at:

    A. T_{10} B. S_2

    C. Micturition center   D. Above T_{10}

190. Dose of pyridoxine in the prevention of neuropathy caused by INH is:

    A. 1 mg   B. 5 mg

    C. 10 mg   D. 50 mg

191. Phenytoin therapy can result in all except:

    A. Ophthalmoplegia

    B. Gingival hyperplasia

    C. Micrpocytic hypochromic anaemia

    D. Ataxia

192. Most common presentation of diabetic neuropathy is:

    A. Mononeuropathy   B. Symmetrical sensory

    C. Amyotrophy polyneuropathy D. Cranial multiple nerve palsies

193. Which of the following is not a malignant brain tumor:

    A. Glioblastoma   B. Medulloblastoma

    C. Haemangioblastoma   D. Ependymoma

194. For the diagnosis and treatment of berry anenrysm, most useful test is:

    A. X-ray skull   B. EEG

    C. Carotid angiography   D. CT scan of brain

195. Not a cause of peripheral neuropathy :

    A. Amyloidosis   B. Sarcoidosis

    C. Leprosy   D. Polyarlteritis nodosa

196. Fasciculations are not seen in :

    A. Myopathy   B. Polymyositis

    C. Motor neurone disease   D. Polio

197. In Brown Sequard syndrome, true is:

    A. No sensory loss   B. Ipsilateral corticospinal signs

    C. Flexor planter reflex   D. Loss of sphincter control

198. Commonest cause of stroke in an elderly is:

    A. Thrombosis   B. Hypertension

    C. Berry aneurysm rupture   D. Head injury

199. Not true about chronic subdural haematoma:

    A. Acute onset of headache

    B. Stroke

    C. Decreased response to stimuli

    D. Always preceded with history of unconsciousness

Ans.: 189. C   190. C   191. C   192. A   193. D   194. C   195. D

      196. A   197. B   198. A   199. D

200. **Alzheimer's disease is associated with:**
   A. Atrophy of frontal and temporal lobes
   B. Atrophy of temporal and parietal lobes
   C. Multiple crabnial nerve palsies
   D. Gemiparesis or hemianesthesia

201. **Tremors may be seen in all of the following except:**
   A. Hyperparathyroidism   B. Acute alcoholism
   C. Hepatic coma   D. Hyperthyroidism

202. **Fasciculations may be seen in :**
   A. Tabes dorsalis
   B. Myopathy
   C. Myoitonic muscular dystrophy
   D. Amyotrophic lateral sclerosis

203. **The most effective treatment for trigeminal neuralgia is:**
   A. Indomethacin   B. Pethidine
   C. Xylocaine   D. Carbamazepine

204. **Familial periodic paralysis is characterized by all of the following except:**
   A. Hyperkalemia   B. Normo or hypoklemia
   C. Adrenaline sensitivity   D. None of the above

205. **Acoustic neuroma is characterized by all except:**
   A. Ataxia   B. Tinnitus
   C. Diplopia   D. Lateral gaze palsy

206. **The complications of trigeminal neuralgia include:**
   A. Convulsion   B. Dryness of eyes
   C. Paresthesias   D. Malnutrition

207. **The aetiology of Bell's palsy is:**
   A. Viral   B. Trauma
   C. Exposure to cold   D. Unknown

208. **The differential diagnosis of Bell's palsy included all of the following except:**
   A. Chronic suppurative otitis media
   B. Cerebral tumor
   C. Mastoiditis
   D. None of the above

209. **Which of the, following is used in the treatment of Bell's palsy:**
   A. Dexamethasone   B. Diazepam
   C. Surgery   D. All of the above

Ans.: 200. A   201. A   202. D   203. D   204. D   205. D   206. D
      207. D   208. D   209. D

210. **Migrainous aura is usually characterized by all of the following except:**
    A. Scintillating scotomas
    B. White or coloured halos
    C. Paraesthesiae or numbness of hands
    D. Olfactory or gustatory hallucinations

211. **Ergotamine, used in the migraine may given:**
    A. Sublingually   B. Orally
    C. Aerosol form   D. In any of the above forms

212. **Oppenheim's gait (Swinging motion of head, body and extremities on walking) may be seen in :**
    A. Syringobulbia   B. Tabes dorsalis
    C. Multiple sclerosis   D. Cerebellar tumor

213. **The following may be used in the treatment of multiple sclerosis:**
    A. Dexamethasone   B. Dantrolene
    C. Baclofen   D. All of the above

214. **The commonest cause of Parkinsonism in children is :**
    A. Postencephalitic   B. Brain tumor
    C. Head injury   D. Drug induced

215. **Which of the following is most useful in the diagnosis of Parkinsonism:**
    A. EEG and CT Scan   B. Clinical examination
    C. History   D. Therapeutic response to drugs

216. **Café-au-lait spots are seen in association with:**
    A. Congenital nystagmus
    B. Mental retardation
    C. Atrophy of distal and proximal musculature
    D. Multiple subcutaneous nodules

217. **Amyotrophie lateral sclerosis is characterized by:**
    A. Jacksonian epilepsy
    B. Cogweel rigidity
    C. Signs of ventral horn and lateral column involvement
    D. Remissions and exacerbations

218. **A subdural hematoma is:**
    A. Always chronic
    B. Rarely seen in children
    C. Rare in the absence of trauma
    D. Always of venous origin

Ans.: 210. D   211. D   212. C   213. D   214. D   215. B   216. D
217. C   218. D

219. A 65 year old woman has episodes lasting up to 4-5 minutes. Which consist of impaired vision in his right eye and numbness of the left side of his body:
- A.   Internal carotid artery insufficiency
- B.   Posterior cerebral artery aneurysm
- C.   Parasagittal meningioma
- D.   Frontal lobe tumor

220. A patient complaining of persistent drowsiness associated with narcolepsy will show:
- A.   Cataplexy   B.   Hypnagogic halluciuations
- C.   Sleep paralysis   D.   All of the above

221. The differential diagnosis of Pakinsonisum includes all except:
- A.   Alcoholism   B.   Heneral paresis of insane
- C.   Multiple sclerosis   D.   Hypothyroidism

222. The differential diagnosis of stroke includes all of the following except:
- A.   Encephalitis   B.   Myoclonic epilepsy
- C.   Multiple scelrosis   D.   Aubdural haematoma

223. Fabere sign (pressing of knee downwards and laterally in patients lying supine with heel of the testing leg placed on the patella of the opposite leg produces limitation of motion due to resistance) is seen in:
- A.   Table dorsalis   B.   Subacute combined degeneration
- C.   Syrinobulbia   D.   Lumbago-Sciatica syndrome

224. The most effective treatment in Lumbago-sciatica syndroma is:
- A.   Bed rest   B.   Hot fomentation
- C.   Paracetamol   D.   Ketazolam

225. Which of the following muscular dystropeies is inherited as autosomal dominant:
- A.   Pseudohypertrophic (Duchenne's)
- B.   Limb-girdle type (Erb's) type
- C.   Facio-scapulo-humoral (Landouzy-Dejerine)
- D.   None of the above

226. All of the are the precipitants of myasthenia gravis except:
- A.   Infection   B.   Pegnancy
- C.   Emotional disturbances   D.   Obesity

227. The differential diagnosis of myasthenia gravis includes all of the following except:
- A.   Poliomyositis   B.   Peripheral neuropathy
- C.   SLE   D.   Cushing's syndrome

Ans.: 219. A   220. D   221. D   222. B   223. D   224. A   225. C
226. D   227. D

228. **The pathologic changes in Friedreich's ataxia are found in all except:**
   A. Posterior funiculi
   B. Spinocerebellar tracts
   C. Lateral corticospinal tract
   D. Basal ganglia

229. **A young patient presents with a history of rapid loss of vision in one eye. Examination reveals pain on movement of the eyeballs. Which of the following may be most useful investigation in establishing the diagnosis:**
   A. X-ray skull
   B. Fundus
   C. Perimetry
   D. Hypnosis

230. **Which of the following is the first symption in herpes zoster:**
   A. Vesicles
   B. Severe pain
   C. Paresthesia
   D. VII Nerve palsy

231. **The commonest demyelinating disease is :**
   A. Post vacinational
   B. Neuromylitis optia
   C. Subacute combined degeneration
   D. Multiple sclerosis

232. **A genetic subsceptiblity to multiple sclerosis is suggested by an increased frequency of all of the following HLA except:**
   A. A_3
   B. B_7
   C. B_{27}
   D. DW2/DRW2

233. **The aetiology of multiple sclerosis is:**
   A. Autoimmune mechanism
   B. Measles infection
   C. Slow viral infection
   D. Unknown

234. **Barber chair sign is seen all of the following except:**
   A. Multiple sclerosis
   B. Syringomyelia
   C. Vitamin B_1 deficiency
   D. Parkinsonism

235. **The immunoglobulins increased in multiple sclerosis is:**
   A. Ig M
   B. Ig G
   C. IgA
   D. IgE

236. **All of the following are the poor prognostie factors in multiple sclerosis except:**
   A. Late teens or early twenties age of onset
   B. Incomplete remission
   C. Early motor, cerebellar or brain stem dysfunction
   D. Initial manifestation is late

237. **Which of the following is an infrequent symptom of acute demyelinating encephalomyelitis :**
   A. Headache
   B. Detirium
   C. Sensory loss
   D. Flaccid paralysis

Ans.: 228. D   229. C   230. B   231. D   232. C   233. D   234. C
   235. B   236. D   237. C

238. **The minimum mortality is seen in which of the following acute demyelinating encephalomyelitis:**
    A. Neuromyelitis optical
    B. Post vaccinational
    C. Postexanthematous
    D. All of the above have equal mortality

239. **Treatment of choice in acute demyelinating encephalomyelitis is:**
    A. ACTH   B. Aspirin
    C. Plasmapheresis   D. Azathioprine

240. **The commonest type pf Parkinsonism with known aetiology is:**
    A. Head injury   B. Encephalitis lethargica
    C. Idiopathic   D. Latrogenic

241. **All of the following are the causes of Parkinson's disease except:**
    A. Manganese poisoning   B. Carbon monoxide poisoning
    C. Wilson's disease   D. Non of the above

242. **In parkinsonism, flexion is seen at all of the following joints except:**
    A. Elbows   B. Hipes
    C. Metacarpophalangeal joints   D. Interphalangeal joints

243. **Which of the following is a common finding in all types of parkinsonism:**
    A. Blephrospasm
    B. Oculogyric crises
    C. Impaired papillary accommodation
    D. Reduction in blinking

244. **All of the following are the differential diagnosis of Parkinsonian tremors except:**
    A. Anxiety   B. Alcoholism
    C. Thyrotoxicosis   D. Cerebellar tremor

245. **Atherosclerotic dementia differs from parkinsonism by all of the following except:**
    A. Pyramidal signs   B. Exaggerated Jaw Jerk
    C. Increased tendon reflexes   D. Spasticity

246. **The average daily dose of levodapa is**
    A. 1 g   B. 2 g
    C. 5 g   D. 8 g

247. **All of the following are inherited as autopsomal dominant except:**
    A. Huntington's chorea   B. Tuberous sclerosis
    C. Myotonic dystrophy   D. Duchenne muscular dystrophy

Ans.: 238. C   239. A   240. D   241. D   242. D   243. C   244. D
        245. A   246. C   247. D

248. **A patient with wilson's disease has the following urinary findings except:**
    A. Aminoaciduria
    B. Phosphaturia
    C. Glycosuria
    D. Bence Jones Proteinuria

249. **Wilson's disease may be treated by:**
    A. Dimercaprol
    B. Disodium calcium versenate
    C. Penicillamine
    D. All of the above

250. **All of the following are the early manifestations of Kernicterus except:**
    A. Opisthotonous
    B. Convulsions
    C. Deafness of nuclear type
    D. Rigidity

251. **Normal value of CSF chloride in mmol/L is:**
    A. 103–115
    B. 115–130
    C. 130–150
    D. 730–750

252. **About Huntington's chorea, all are true except:**
    A. Autosomal dominant with incomplete penetrance
    B. Intellectual deteriroration
    C. Chorea
    D. Young people are affected and has a rapid deteriorating course

253. **Foot drop is seen injury to:**
    A. Common peroaeal nerve
    B. Tibial nerve
    C A chillis tendon
    D. Popliteal nerve

254. **Commonly seen in kernicterus:**
    A. Hydrocephalus
    B. Athetosis
    C. Brown staining of teeth
    D. Mental retardation

255. **Drug contraindicated in myasthenia gravis is:**
    A. Ephedrine
    B. Edrophonium
    C. Quinine
    D. Pyridostigmine

256. **Meralgia paraesthetia involves:**
    A. Axillary nerve
    B. Sural nerve
    C. Median nerve
    D. Lateral cutaneous nerve of thigh

257. **The commonest primary brain tumor is:**
    A. Astrocytoma
    B. Glioblastoma
    C. Meningioma
    D. Metastasis

258. **Drug best in controlling the movements of Huntington's chorea is:**
    A. Haloperidol
    B. Diazepam
    C. Tetrabenazine
    D. Phenytoin

Ans.: 248. D   249. D   250. C   251. B   252. A   253. A   254. A
255. C   256. D   257. A   258. C

259. **Dementia associated with Huntington's chorea is:**
   A. Progressive
   B. Transient
   C. Waxing and Waning
   D. Reversible with the control of choreiform movements

260. **Rapid, repetitive, coordinated and stereotyped movements, most of which can be mimicked are known as:**
   A. Chorea　　　　　　　　　B. Ballism
   C. Athetosia　　　　　　　　D. Ties

261. **Athetoid movements are most marked in the :**
   A. Face
   B. Trunk
   C. Proximal part of extremities
   D. Distal part of extremities

262. **Following are the actiologies implicated in the causation of motor neurone disease:**
   A. Aluminium poisoning　　　B. Lead poisoning
   C. Autommune phenomenon　 D. Slowviral diseases

263. **The earliest feature of progressive bulbar palay is:**
   A. Dysphagia　　　　　　　　B. Hoarsness of voice
   C. Impaired articulation　　　　D. Wasting of tongue

264. **The commonest presentation of patients with progressive muscular atrophy is:**
   A. Wasting and weakness of small muscles of hands
   B. Spasticity of the legs
   C. Foot drop
   D. Dysphagia

265. **The only sensory symptoms present in motor neuroen disease are related to:**
   A. Pain　　　　　　　　　　B. Touch
   C. Vibration　　　　　　　　D. Position

266. **The differential diagnosis of motor neurone disease include all except:**
   A. Diabetic amyotrophy　　　　B. Meningovascular syphilis
   C. Ocuit carcinoma of bronchus D. Subacute combined degeneration

267. **The motor neurone disease type with best prognosis is:**
   A. Progressive bulbar palsy
   B. Amyotrophic lateral sclerosis
   C. Progressive muscular atrophy
   D. All have same prognosis

Ans.: 259. A　260. D　261. D　262. B　263. C　264. C　265. A
　　　266. D　267. C

268. The unaffected members of the family having a patient with Friedreich's ataxia may show:
   - A. Renal defects
   - B. Pes cavus
   - C. Retrobulbar neuritis
   - D. Ventricular septal defect

269. In syringomyelia , dissociated sensory loss is related to:
   - A. Pain and temperature
   - B. Touch
   - C. Position
   - D. Vibration

270. The cranial nerve which is most commonly involved with von Recklinghausen's disease (Neurofibromatosis) is:
   - A. II
   - B. III
   - C. VI
   - D. VIII

271. The treatment of choice for pain associated with Syringobulbia is:
   - A. Morphine
   - B. Chlorpromazine
   - C. Surgery
   - D. Radiotherapy

272. The isolated deficiency of Pyridoxine may occur in infants in whom the common presentation is:
   - A. Dysarthria
   - B. Diarrhoea
   - C. Delayed milestones
   - D. Convulsions

273. In alcoholics, the pathological lesions may be found in all of the following except:
   - A. Midbrain
   - B. Pons
   - C. Cerebellum
   - D. Hypothalamus

274. Which of the following symptoms is not seen in subacute degeneration of cord:
   - A. Glove and stocking impairment of superficial sensation
   - B. Position and vibration sense markedly impaired in upper limbs
   - C. Ankle jerks absent
   - D. Extensor plantars

275. The daily dose of hydroxocobalamine in subacute combined degeneration is:
   - A. 10 p.g.
   - B. 100 p.g.
   - C. 1000 p.g.
   - D. 5000 p.g.

276. In syringomyelia, Charcot Joints are common in :
   - A. Upper limbs
   - B. Lower limbs
   - C. Spine
   - D. None of the above

277. All of the following are true about burning feet syndrome except:
   - A. Mild but painful sensory neuropathy occurring in elderly
   - B. Severe lancinating or burning pains in the feet and distal parts of the leg
   - C. Pain is severe in bed at night
   - D. Signs of polyneuropathy are often present

Ans.: 268. B 269. A 270. D 271. D 272. D 273. D 274. B
275. C 276. A 277. D

278. **A space occupying lesion within the spinal canal may involve nerve issue by:**
    A. Pressure
    B. Interfering with blood supply
    C. Oedena from venous obstruction
    D. All of the above

279. **The three major causes of cord compression include all of the following except:**
    A. Tumors (extramedullary)   B. Disc prolapse
    C. Infections   D. Trauma

280. **All of the following are common bladder symptoms of cord compression except:**
    A. Urgency   B. Hesitancy of micturition
    C. Urinary retention   D. Urinary incontinence

281. **Lesions above which cervical segment lead of tetraplegia:**
    A. C_5 B. C_6
    C. C_7 D. C_8

282. **Froin's syndrome is characterized by all except:**
    A. Xanthochromia   B. Increase in protein content
    C. Ieucocytosis   D. None of the above

283. **The commonest presentation of acute herniation of disc is:**
    A. Hyperalgesia + Hyperanaesthesia
    B. Paraesthesia
    C. Depression of tendon reflexes
    D. Lower motor neurone paresis in root distribution

284. **In cervical spondylosis, there is:**
    A. Increase in kyphosis   B. Decrease in kyphosis
    C. Increase in lordosis   D. Decrease in lordosis

285. **Lumbago is defined as the pain in the :**
    A. Lower part of the back
    B. Lower part of the back+in the distribution of sciatic nerve
    C. Distribution of sciatic nerve
    D. Any where in the back

286. **The commonest presentation of the Lumbago- Sciatica syndrome is:**
    A. Loss of ankle jerk
    B. Foot drop
    C. Weakness of inversion of the foot
    D. Loss of Knee jerk

Ans.: 278. D   279. C   280. D   281. A   282. C   283. A   284. B
285. A   286. A

287. **Which of the following is not true about carpal tunnel syndrome:**
    A. Most frequent in middle aged females
    B. If it occurs in pregnancy, thiazides may be the treatment
    C. The common complaints are pain' numbness and tingling sensation and are often bilateral
    D. Objective sensory loss of radial one and a half digits

288. **Which of the following is associated with aggravation of pain of carpal tunnel syndrome:**
    A. Puerperium   B. Sleep
    C. Using of fingers   D. All of the above

289. **Lasegue's sign is most characteristically seen in:**
    A. Cervical spondylosis
    B. Lumbago-sciatica syndrome
    C. Rheumatoid arthritis in cervical spine
    D. Subacute combined degeneration

290. **Which of the following is not a cause of toxic polyneuropathy**
    A. Lead   B. Arsenic
    C. Manganese   D. Mercury

291. **All the following are the causes if netabolic neuropathies except:**
    A. Renal failure   B. Hepatic failure
    C. Acute intermittent porphyria D. None of the above

292. **The increase in CSF proteins in the Guillain Barre syndrome may not be observed during first :**
    A. 24 hours   B. 72 hours
    C. 10 days   D. 21 days

293. **Which of the following is most important aspect of management of the Guillain-Barre syndrome:**
    A. Steroids   B. Muscle relaxants
    C. Physiotherapy   D. Maintence of respiration

294. **The following drugs are contraindicated in acute intermittent porphyria except:**
    A. Pentazocine   B. Methyldopa
    C. ACTH   D. Chlorpropamide

295. **The following illnesses may be precipitated by pregnancy except:**
    A. Carpal tunnel syndrome
    B. Myasthenia gravis
    C. Tropical sprue (folate deficiency)
    D. Neuropathy by saroidosis

Ans.: 287. D   288. B   289. B   290. C   291. D   292. C   293. D
          294. C   295. D

296. **The following infections diseases are associated with neuropathy except:**
   A. Influenza   B. Measles
   C. Diphtheria   D. Mumps

297. **Which the following joint is most commonly affected by myasthenia gravis:**
   A. Pelvis   B. Elbow
   C. Shoulder joint   D. Sternoclavicular

298. **The most common presentation of encephalopathy associated with carcinoma of bronchus; uterus or breast is:**
   A. Dementia   B. Depression
   C. Delirium   D. Anxiety neurosis

299. **All of the following drugs may produce myopathy except:**
   A. Propranolol   B. Cimetidine
   C. Lithium   D. Guanidine

300. **The myopathy caused by corticosteroids or Cushing's syndrome most commonly affect joint:**
   A. Shoulder   B. Pelvis
   C. Knee   D. Wrist

301. **Pseudodementia, in which the patient often complains of forgetfulness, is seen in:**
   A. Brain infarction   B. Epilepsy
   C. Alzheimer's disease   D. Depression

302. **In Alzheimer's disease, most commonly affected function of speech is:**
   A. Naming an object   B. Comprehension
   C. Fluency   D. General information

303. **Lesion in Broca's area may lead to:**
   A. Fluent aphasia and hemiplegia
   B. Non-fluent aphasia and hemiplegia
   C. Fluent aphasia only
   D. Fluent or non-fluent aphasia

304. **First neurotransmitter discovered was:**
   A. Norepinephrine   B. Dopamine
   C. Serotonin   D. Acetylcholine

305. **Gertsman syndrome is seen in lesion of:**
   A. Frontal lobe   B. Parietal lobe
   C. Temporal lobe   D. Occipital lobe

Ans.: 296. D   297. C   298. A   299. D   300. B   301. D   302. A
303. B   304. D   305. B

306. **Match the following:**

(I)	Frontal lobe	(i)	Visual ionattention
(II)	Parietal lobe	(II)	Grasp reflex
(III)	Temporal lobe	(III)	Anosognosia
(IV)	Occipital lobe	(IV)	Kluver Bucy syndrome

    A. I (ii), II(iv), III (iii), IV(i)
    B. I(iv), II(ii), III(iii), IV(i)
    C. I(ii), II(iii), III(iv), IV(I)
    D. I(ii), II(iv), III(iii), IV(i)

307. **Which of the following disturbances in psychomotor activity is pathognomic of delirium :**
    A. Stereotypy
    B. Echopraxia
    C. Waxy flexibilities
    D. Multifocal myoclonus

308. **Most frequent cause of amnestic syndrome is:**
    A. Thiamine deficiency
    B. Head injury
    C. Stroke
    D. Ebcephalitis

309. **Which of the following is an uncommon symptom of dementia:**
    A. Memory impppairment
    B. Impairement in consciousness
    C. Impaired abstract thinking
    D. Personality change

310. **Which of the following is an untreatable form of dementia:**
    A. Hypothyroidism
    B. Addison's disease
    C. Alcoholic encephalopathy
    D. Normal pressure hydrocephalus

311. **A cerebral activator, piracetam is absolutely contraindicated in:**
    A. Coronary artery disease
    B. Renal dysfunction
    C. Psychoses
    D. Pregnancy

312. **Co-dergocrine mesylate, a cerebral activator is contraindicated in:**
    A. Epilepsy
    B. Psychoses
    C. Peptic nicer
    D. All of the above

313. **Levodopa is contraindicated in all except:**
    A. Schizophrenia
    B. Malignant melanoma
    C. Narrow angle glaucoma
    D. Peptic ulcer

314. **Levodopa is contraindicated along with use of:**
    A. Trihexyphenidyl
    B. Imipramine
    C. MAO inhibitors
    D. Amantadine

Ans.: 306. C   307. D   308. B   309. B   310. C   311. D   312. B
     313. D   314. C

315. **Bromocriptine is used in the treatment of all of the following except:**
    A. Postencephalitic Parkinson's disease
    B. Hyperprolactinemia
    C. Acromegaly
    D. None of the above

316. **Migrainous headaches may be associated with all of the following except:**
    A. Dysphasia   B. Diplopia
    C. Paresthesia   D. Seizure

317. **The diagnosis in an infant with choreoretinitis, epileptic fits and evidence of calcification in the brain is most likely to be suffering from:**
    A. Toxoplasmosis   B. Congenital syphilis
    C. Kernicterus   D. Wilson's disease

318. **Huntington's chorea is characterized by all of the following except:**
    A. Autosomal dominant with complete penetrance
    B. Dementia
    C. Atrophy of cortex and enlagement of ventricles
    D. Onset in childhood

319. **The cardinal feature of Parkinson's disease is:**
    A. Constant fine tremor   B. Muscle dystrophy
    C. Stooped posture   D. Akathisia

320. **Ophthalmoplegia may be found in:**
    A. Parkinson's disease   B. Ophthalmic zoster
    C. Syrinomyelia   D. Migraine

321. **The cerebral function starts impairing when total cerebral blood flow falls below:**
    A. 70%   B. 60%
    C. 55%   D. 30%

322. **A patient with motor dysphasia, apraxia, disturbance in micturition and hemiplegia in which the leg is weaker than the arm is most likely to have infarction in the area supplied by:**
    A   Anterior cerebral artery B. Middle cerebral artery
    C. Posterior cerebral artery D. Posterior communicating artery

323. **Brain stem infarction is characterized by all of the following except:**
    A. Crossed paralysis and nystagmus
    B. Ipsilateral one or multiple cranical nerve lesions
    C. Contralateral pyramidal signs
    D. None of the above

Ans.: 315. D   316. D   317. A   318. D   319. C   320. D   321. C
      322. A   323. D

324. If the symptoms produced by transient ischaemic attach last more than what duration, the infarction has probably occurred;
    A. 15 minutes
    B. 30 minutes
    C. 45 minutes
    D. 60 minutes

325. A significant difference between the blood pressures measured in each arm and a bruit audible over the supraclavicular fossa on the affected side indicates:
    A. Coarctation of aorta
    B. Thalamic syndrome
    C. Subclavian steal syndrome
    D. Cartoid artery stenosis

326. Which of the following disorder may mimic Parkinson's disease:
    A. Meningococcal meningitis
    B. Pseudobullar palsy
    C. Wilson's disease
    D. Posterior artery occlusion

327. The commonest source of cerebral emboli is:
    A. Rheumatic valvular lesions
    B. Cardiomyopathy
    C. Infective endocarditis
    D. Old myocardial infarction

328. Isolated third cranial nerve palsy may be seen in :
    A. Middle cerebral artery aneurysm
    B. Anterior cerebral artery aneurysm
    C. Posterior communicating artery aneyrysm
    D. Posterior cerebral artery aneurysm

329. Lenticulostriate artery in internal capsule (also known as artery of hemorrhage) is a branch of:
    A. Anterior cerebral artery
    B. Middle cerebral artery
    C. Anterior choroidal artery
    D. Posteriar cerebral artery

330. Haemorrhage into the pons is characterized by all of the following except:
    A. Periodic respiration
    B. Hyperpyrexia
    C. Unilateral involvement of cranial nerves
    D. Bilateral pyramidal signs

331. The first manifestation of an intracranial aneurysm is often:
    A. Nerve palsy
    B. Subarachnoid haemorrhage
    C. Papilloedema
    D. Hemiparesis

332. Caroticocavernous fistula is characterized by all of the following except:
    A. Pulsatile exophthalmos
    B. Paresthesias
    C. Papilloedema
    D. Complete ophthalmoplegia

Ans.: 324. D   325. C   326. B   327. D   328. C   329. B   330. C
331. B   332. B

333. **Subarachnoid haemorrhage may be associated with all of the following except:**
    A. Severe and sudden headache   B. Hemiparesis
    C. Aphonia   D. Diplopia

334. **Lucid interval is seen in :**
    A. Venous haemorrhage from subdural space
    B. Middle meningeal artery rupture
    C. Middle cerebral artery aneurysm
    D. Subarachnoid haemorrhage

335. **Transient ischaemic attack, in contrast to migraine, is characterized by all of the following except:**
    A. Consciousness unimpaired
    B. Non-throbbing headache
    C. Onset after the age of 10 years
    D. None of the above

336. **In echoencephalography, a shift of the mid-line echo most often suggests:**
    A. Haemorrhage   B. Infarction
    C. Aneurysm   D. Thrombosis

337. **Which of the following is ineffective in the treatment of an established stroke:**
    A. Low molecular weight dextran
    B. Dexanethasone
    C. Salicylates
    D. Streptokinase

338. **Anticoagulants are most useful in:**
    A. Transient ischaemic attack   B. Established stroke
    C. Stroke in evolution   D. Cerebral haemorrhage

339. **Which is the most frequent lesion in an elderly in a clinical practice:**
    A. Cerebral aneurysm   B. Cerebral infarction
    C. Cerebral haemorrhage   D. Brain tumors

340. **All of the following are the causes of aseptic cerebral thrombisis except:**
    A. Polycythaemia
    B. Malaria
    C. Water and salt depletion
    D. Idiopathic thrombocytopenic purpura

341. **The commonest symptom of superior sagittal sinus thrombosis:**
    A. Severe headache   B. Hemianopia and aphasia
    C. Hemiplegia   D. Convulsion

Ans.: 333. C   334. B   335. B   336. A   337. D   338. C   339. B
      340. D   341. A

342. Which of the following symptoms indicates the extension of thrombosis from transverse sinus to superior sagittal sinus:
  A. Faciobrachial paresis
  B. Hemiparesis
  C. Paraparesis
  D. Hemianopia

343. Which of the following symptoms is often common in thrombosis of superior sagittal, transverse and cavernous sinuses:
  A. Headache
  B. Cranial nerve palsy
  C. Hemiplegia
  D. Hemianopia

344. The following symptoms are charateristic of cavernous sinus thrombosis except:
  A. Headache
  B. III, IV and VI nerve palsies
  C. Papilloedema in the affected eye
  D. Lynphocytic leucocytosis

345. Antibiotics are most useful in the thrombosis in :
  A. Superior sagittal sinus
  B. Cavernous sinus
  C. Transverse sinus
  D. Cortical veins

346. The most useful investigation in tubercular meningitis is:
  A. Guinea pig inoculation
  B. Culture of CSF
  C. Smear from CSF
  D. Mantoux test

347. Which of the following may never revert to normal in the treatment of neurosyphilis:
  A. Increeased cells
  B. Elevated protein
  C. Bacteraemia
  D. Wasserman reaction

348. Tabetic facies is typically characterized by all of the following except:
  A. Bilateral ptosis
  B. Compensatory wrinkling
  C. Srgyll Robertson pupil
  D. Defects in external ocular movements

349. The first tendon reflex lost ion tabes dorsalis is:
  A. Knee
  B. Ankle
  C. Biceps
  D. Supinaor

350. Subacute sclerosing panencephalitis is characterized by all of the following except:
  A. Sequence of infection with myxovirus
  B. Occurs in children and adolescents
  C. Grand mal seizures common
  D. EEG shows triphasic slow waves

Ans.: 342. C   343. A   344. D   345. B   346. A   347. D   348. D
349. B   350. C

351. **All are ture about electroencephalography except:**
   A. Is more effective in detecting gliomas than meningiomas
   B. In acute lesions, the changes are marked than chronic lesions
   C. Is the most effective wway of deciding whether patient is epileptic or not.
   D. Is more precise in its definition of hemishere than of posterior fossa lesions.

352. **Which of the following is most useful signs of brain death:**
   A. Absent corneal reflexes
   B. Fixed dilated pupils not responding to light
   C. The absence of spontaneous respiration even when pCO, is greater than 50 mm Hg
   D. The absence of electrical activity of cerebral origin on the EEG

353. **Which of the following cerebrospinal fluid (CSF) contains minimal protein:**
   A. Lumbar   B. Cisternal
   C. Ventricular   D. Thoracic

354. **Tibulant atxia (ataxia with vertical oscillation of head, trunk and arm) is typically seen in :**
   A. Posterior column lesion   B. Peroneal muscular atrophy
   C. Disseminated sclerosis   D. Themors of vermis

355. **Pseudoparkinsonism is caused by:**
   A. Haloperidol   B. Reserpine
   C. Tebrabenazine   D. All of the above

356. **Choreiform movements may be induced by all except:**
   A. Phenytoin   B. Amphetamine
   C. Haloperidol   D. Alprasolam

357. **Very slow, irregular, twisting movement, one movement perpertually blending with another is called:**
   A. Ballism   B. Chorea
   C. Athetosis   D. Tremors

358. **Flapping tremor may occur in all except:**
   A. Hepatic precoma   B. Uremia
   C. Respiratory failure   D. Wilson's disease

359. **Match the following**

Tremor type		Disease
I.	Intention tremor	(i) Wilson's Disease
II.	Wing beating tremor	(ii) Barbiturate intoxication

Ans.: 351. C   352. D   353. C   354. C   355. D   356. D   357. C
358. D   359. C

III. Searching tremor   (iii) Disseminated sclerosis

IV. Flapping tremor   (iv) Parietal lobe lesion

A. I (i)   II (ii)   III (iii)   IV (iv)

B. I (ii)   II (iii)   III (i)   IV (iv)

C. I (iii)   II (i)   III (i)   IV (iv)

D. I (ii)   II (i)   III (iii)   IV (iv)

360. **Match the following :**

Bladder   *Lesion in*

I. Uninhibited   (i) Posterior Column column

II. Automatic (Reflex)   (ii) Conus or Cauda

III. Autonomous   (iii) Spinal cord above sacral segment

IV. Sensory atonic   (iv) Motor leg area internal capsul

A. I (i)   II (ii)   III (iii)   IV (iv)

B. I (ii)   II (i)   III (iv)   IV (iii)

C. I (iv)   II (iv)   III (ii)   IV (i)

D. I (iii)   II (iv)   III (i)   IV (ii)

361. **An extensor plantar response can only occur in lesion above:**

A. T_{11} vertebra   B. T_{12} vertebra

C. L_1 vertiebra   D. L_2 vertebra

362. **Babinski sign may be most commonly detected is:**

A. Pyramidal tract lesion   B. Epilepsy

C. Sleep   D. Hypogycemia

363. **All are features of Brown Sequard syndrome except:**

A   Babinski sign positive indicates the side of the lesion

B. Loss of proprioception in the leg on the side of lesion

C. A band of hyperaesthesia at the level of lesion on the opposite side

D. Loss of temperature sensation in the leg opposite to the lesion

364. **A patient has extensor plantar responses and wasting of the small muscle of the hand, He has:**

A. Guillain barre syndrome

B. General paralysis of insane

C. Motor neurone disease

D. Progressive muscular dystroyhy

365. **All are common causes of subcortical dementia except:**

A. Progressive supranuclear palsy (Steel Richardson syndrome)

B. Parkinson's disease

C. Creutfeldt Jacob's disease

D. Wilson's disease

Ans.: 360. C   361. B   362. C   363. C   364. C   365. C

366. Which of the following drugs, if given alone, is useful in the treatment of parkinsonism :
    A. Alpa-methyldopa
    B. Carbidopa
    C. Monoamine oxidase inhibitors
    D. Bromocriptine

367. The following are the recognized manifestation of brainstem infarction except:
    A. Diplopia   B. Nystagmus
    C. Motor dysphasia   D. Severe headache

368. All of the following are poor prognostic factors in stroke except:
    A. Dementia
    B. Uncontrolled hypertension
    C. Episodes of unconsciousness
    D. Embolic type rather than infarction or hemorrhage type

369. Dilatation of arteries (intracranial extracranial) is thought to be mechanism responsible for headache caused by:
    A. Post herpetic neuralgia   B. Meningitis or encephalitis
    C. Lumbar puncture   D. Chronic respiratory failure

370. Which of the following is true about carpal tunnel syndrome:
    A. Remits during Pregnancy
    B. Cause wasting of interosseus and lumbrical muscles
    C. Thyrotoxicosis may complicate it
    D. May interfere with sleep

371. A brain tumor with a tendency to metastasize to other areas of cerebrum and spinal cord is:
    A. Meningioma   B. Medullloblastoma
    C. Ependymoma   D. Craniopharyngioma

372. Inspiratory stridor is seen in:
    A. Unlateral total paralysis of cord
    B. Unlateral abductor paralysis
    C. Bilateral abductor paralysis
    D. Bilateral total paralysis

373. Lesion in the superior temporal gyrus results in:
    A. "Pure" word blindness   B. "Pure" word muteness
    C. "Pure' word deafness   D. Transcortical aphasia

374. Occlusion of lower division of middle cerebral artery on the dominant side is characterized by:
    A. Broca's aphasia   B. Wernicke's aphasia
    C. Amnestic dysnomic aphasia   D. Global aphasia

Ans.: 366. D   367. C   368. D   369. D   370. D   371. B   372. C
        373. C   374. B

375. Memory is stored in :
   A. Frontal lobe   B. Parietal lobe
   C. Hippocampus   D. Hypothalamus

376. Prolonged ankle jerks are seen in all except:
   A. Myasthenia gravis   B. Parkinsonism
   C. Hypokalemia   D. Hypeerthyroidism

377. Locked – in – syndrome may be seen in all expect:
   A. Pontine lesions
   B. Tegmentum lesions
   C. Bilateral lesions of medulla
   D. Bilateral lesions of lateral one third of cerebellar peduncles

378. Extensor plantar response will not be detected even in the presence of a pyramidal lesion if there is presence of all of the following except:
   A. Gross loss of senoation on the sole of the foot
   B. Hallux regidus
   C. Paralysis of extensor hallucis longus
   D. Paralysis of extensor hallucis brevis

379. Inverted supinator jerk indicates lesion at:
   A. C_4C_5 B. C_5C_6
   C. C_7C_8 D. C_8C_1

380. Bilateral symmetrical brisk reflexes may be seen in:
   A. Pyramidal tract lesion   B. Hysteria
   C. Neurasthenia   D. All of the above

381. The conversion of extensor plantar response in infancy to flexor plantar type occurs usually in the age group of:
   A. 3-6 months   B. 6-9 months
   C. 9-12 months   D. 20-24months

382. All are causes of Argyll- Robertson pupil except:
   A. Syringomyelia   B. Disseminated sclerosis
   C. Chronic alcoholism   D. Asphyxia and deep anaesthesia

383. Which of the following is a cause of bilateral hypoglossal nerve paralysis
   A. Syringobulbia   B. Head injury
   C. Aneurysm of vertebral artery D. Progressive bulbar paralysis

384. Long term complications of phenytoin intake include all except:
   A. Ataxia
   B. Gum hypertrophy
   C. Megaloblastic anemia due to B_{12} deficiency
   D. Osteomalacia

Ans.: 375. C   376. D   377. D   378. D   379. B   380. D   381. C
     382. D   383. D   384. C

385. Transient attacks of dizziness, paraesthesias, diplopia and slurred speech suggest the diagnosis of:
   A. Hypertensive encephalopathy
   B. Occlusion of superior division of middle cerebral artery
   C. Pseudobulbar palsy
   D. Vertebrobasilar insufficiency

386. Migraine may be associated with all of the following except:
   A. Paraesthesia   B. Dysphagia
   C. Seizures   D. Diplopia

387. Extradural hemorrhage is generally due to:
   A. Anticoagulants intake
   B. Rural venous sinus rupture
   C. Brain tumor
   D. Bleeding from middle meningeal artery

388. Most common region of spinal cord to become involved with Hodgkin's disease is:
   A. Cervival   B. Thoracic
   C. Lumbar   D. Sacral

389. Intervertebral disc rupture with signs of cord compression is best treated by:
   A. Strict bed rest   B. Muscle relaxants
   C. Leg traction   D. Immediate surgery

390. Which of the following supports the correct diagnosis of paralysis agitans:
   A. Oculogyric crises
   B. Slow monotonous speech
   C. Loss of emotional control
   D. Bilateral extensor plantar

391. All are true about neurosyphilis except:
   A. Intra-muscular procaine benzeylpenicillin is the treatment of choice
   B. Near all patients with active disease show high cell counts in the CSF
   C. The raised CSF cell count comes to normal as a result of effective therapy
   D. Treatment of neyrosyphilis should be continued till the CSF Wasserman reaction is negative

392. Argyll Robertson pupil is characterized by all except:
   A. Reaction to convergence   B. Irregularity
   C. Failure to react to light   D. Wide dilatation

Ans.: 385. D   386. C   387. D   388. B   389. D   390. B   391. D
      392. D

393. **A low glucose level in CSF is compatible with the diagnosis of:**
   A. Viral meningitis   B. Fungal meningitis
   C. Carcinomatous meningitis   D. Poliomyelitis

394. **Syndrome of familial periodic paralysis may be associated with:**
   A. Hypokalemia   B. Normokalemia
   C. Hyperkalemia   D. All of the above

395. **In CSF, all the following ions are in equal concentration to that of plasma except:**
   A. Sodium   B. Potassium
   C. Chloride   D. Urea

396. **Which of the following finding is abnormal in CSF:**
   A. Clear watery fluid
   B. Glucose and protein are lower as compared to serum beds
   C. Salicylates and thiocyanates normality don't cross blood brain barrier
   D. 2-5 polymorphs/cmm

397. **The commonest type of hydrocephalus :**
   A. Compensatory   B. Obstructure
   C. Otitic   D. Communicating

398. **The type of hydrocephalus seen in dementia is:**
   A. Communicating   B. Otitic
   C. Congenital obstructive   D. Compensatory

399. **Benda's sign (a dolicocephalic head with a bulging occiput) is diagnostic of**
   A. Communicating hydrocephalus
   B. Otitic hydrocephalus
   C. Obstructive at aqueductal site
   D. Obstructive at cisterna magna

400. **Akinetic mutism is seen in all of the following except:**
   A. Bilateral frontal lobe lesions   B. Diffuse cortital damage
   C. Cerebellar lesion   D. Midbrain lesion

401. **In plain X-ray, inion at a higher level from 2nd cervical spine (Taggart and Walker's sign) differenciates:**
   A. Aqueductal and 4th ventricular foramina block
   B. Compensatory and otitic hydrocephalus
   C. Otitie and communicating hydrocephalus
   D. Acute and chronic hydrocephalus

402. **The commonest site of haemangioblastoma is:**
   A. Cerebrum   B. IV Ventricle
   C. Cerebellum   D. Pons

Ans.: 393. C   394. D   395. C   396. D   397. B   398. D   399. C
    400. D   401. A   402. C

403. Which of the following brain tumor is known as **Rathke's Pouch Tumors (rich in cholesterol)**
   A. Chromophobe adenoma   B. Eosinophilie adenoma
   C. Craniopharyngioma   D. Medulloblastoma

404. Commonest site of tuberculoma in children is:
   A. Parietal cortex   B. Temporal cortex
   C. Midbrain   D. Cerebellum

405. Vomiting is an uncommon symptom in brain tumor from:
   A. Temporal lobe   B. III ventricle
   C. IV ventricle   D. Hypothalamus

406. The site of Todd's paralysis is most useful in determining :
   A. Aetiology of convulsion B. Type of seizure
   C. Localization of lesion   D. Prognosis of seizure

407. The most common symptom of corpus collasum lesion is:
   A. Motor apraxia   B. Sensory apraxia
   C. Alexia   D. Agrahia and astereognosia

408. In vermis lesion, the commonest presentation is:
   A. Nystagmus   B. Past pointing sign
   C. Adiadochokinesis   D. Truncal ataxia

409. Foster Kennedy syndrome (papilloedema is one eye and optic atrophy in the other) is:
   A. Medulloblastoma in cerebellem
   B. Tuberculoma in temporal lobe
   C. Meningioma in olfactory groove
   D. Corpus collasum tumor

410. Which of the following symptoms of cerebellar region may also be seen in frontal lobe lesions:
   A. Truncal ataxia   B. Nystagmus
   C. Past pointing sign   D. Intention tremor

411. Earliest sign of raised intracranial pressure in children is:
   A. Erosion of anterior clinoid process
   B. Erosion of posterior clinoid process
   C. Suture separation
   D. "Finger printing or "Silver beaten" appearance

412. Earliest sign of raised intracranial pressure in adults is:
   A. Erosion of anterior clinoid process
   B. Erosion of posterior clinoid process
   C. Enlargement of sella
   D. Silver beaten appearance

Ans.: 403. C   404. D   405. C   406. C   407. A   408. D   409. C
      410. D   411. D   412. B

413. Echoencephalography is most useful in the diagnosis of intracranial:
  A. Abscess
  B. Tumors
  C. Haematomas
  D. Hydrocephalus

414. Colour brain scanning is done by using:
  A. Conray
  B. Mercury
  C. Technetium
  D. Phosphorus

415. The commonest type of meningitis is:
  A. Pachymeningitis cervicalis hypertrophiica
  B. Pachymeningitis haemorrhagica interna
  C. Leptomeningitis
  D. Meningism

416. The commonest reason of low chloride level in CSF in tubercular meningitis is:
  A. Vomiting
  B. Dehydration
  C. Sweating
  D. Diarrhoea

417. "Bromide-Partition test", the serum and CSF bromide (radioactive) ratio less than 1.6 quickly differentiates:
  A. Pyogenic meningitis from tubercular type
  B. Tuberculous meningitis from fungal type
  C. Tuberculous meningitis from viral type
  D. Viral encephalitis from slow viral disease

418. Convulsions may occasionally be a side effect of antitubercular drug:
  A. Rifampicin
  B. INH
  C. Pyrasinamide
  D. PAS

419. Commonest type of meningitis seen in hospitals is:
  A. Meningococcal
  B. Tuberculous
  C. Aseptic (acutue Iymphoctyic choriomeningitis)
  D. Fungal

420. CSF- blood ratio of glucose is:
  A. 1.0
  B. 2.0
  C. 0.5
  D. 0.75

421. Syphilitic meningitis most common affects nerve:
  A. II
  B. III
  C. VI
  D. VII

422. Tromhone tremor in tongue is seen in:
  A. Cerebellar lesion
  B. Anxiety
  C. Lithium toxicity
  D. General paralysis of insane

Ans.: 413. C 414. C 415. C 416. C 417. C 418. B 419. C
  420. C 421. B 422. D

423. **The commonest presentation of Lissauer's type of GPI is:**
    A. Hemiplegia
    B. Loss of tendon reflexes
    C. Urinary incontinence
    D. Fainting or epileptic spells

424. **Lighting pain, once considered pathognomic of tabes dorsalis, is also seen in:**
    A. Tuberculous meningitis
    B. Parietal lobe tumors
    C. Subacute combined degeneration
    D. Diabetic peripheral neuropathy

425. **First favourable change in CSF in the treatment of neurosyphilis is:**
    A. Fall of cell count
    B. Fall in protein content
    C. VDRL test becomes negative
    D. Fall of gamma globutins

426. **Treatment of choice of the "therapeutic paradox" of stokes (Herxheimer raction) in the management of neurosyphitis is:**
    A. Antihistaminics
    B. Prednisolone
    C. Bismnth subsalicylate
    D. Penicillin

427. **The Organism used in the treatment of general paralysis of insane by malaria therapy is:**
    A. Plasmodium ovale
    B. Plasmodium vivax
    C. Plasmodium malariae
    D. Plasmodium falciparium

428. **Which of the following is a contraindication for malaria theraphy in neurosyphillis :**
    A. Tabes doralis
    B. Primary optic atrophy
    C. Cardiovascular lesions
    D. General paralysis of insane

429. **Earliest finding of subacute sclerosing panencephalitis in a child is:**
    A. Alteration in behaviour
    B. Myoclonic jerks
    C. Overt dementia
    D. Stupor

430. **Occlusion of anterior cerebral artery proximal to anterior communicating artery frequently presents as:**
    A. Ipsilateral Sensory – motor paralysis
    B. Contralateral sensory – motor paralysis
    C. Gait disturbances
    D. Asymptomatically

431. **Which of the following drugs does not has anti-platelet adhesiveness effect but is a vasodilator:**
    A. Aspirin
    B. Dipyrldamole
    C. Clofibrate
    C. Betahistine

Ans.: 423. D   424. D   425. A   426. C   427. B   428. C   429. A
    430. D   431. D

432. "Uncinate fits" is a form of:
    A. Illusion   B. Hallucination
    C. Ideokinetic apraxia   D. Reflex epilepsy

433. Lhermite's sign (an abnormal sensation like electric current or vibration, most commonly referred down the back is produced by flexion of the neck and disappears on repeated flexion) is seen in :
    A. Spasmodic torticollis   B. Occipital lobe tumors
    C. Erb's spinal syphilis   D. Multiple sclerosis

434. Match the following :

I.	Carswell and Cruveilhier	(i)	Diagnosis of multiple sclerosis
II.	Von Economo	(ii)	Slow virus infection
III.	Sigurdsson	(iii)	Recognized multiple sclerosis
IV.	Frerich	(iv)	Encephalilis lethargica

    A. I (i)   II (iii)   III (ii) IV (iv)
    B. I (ii)   II (iii)   III (iv) IV (i)
    C. I (iii) II (iv)   III (ii) IV (i)
    D. I (iv) II (iii) III (ii) IV (i)

435. Which of the following mental changes is an early symptom of multiple sclerosis:
    A. Dementia   B. Euphoria
    C. Depression   D. Anxiety

436. All of the following are characteristics of primary optic atrophy except:
    A. Disc margins are well marked
    B. Optic cup is full
    C. Is always secondary to papilloedema
    D. Arteries are constricted and veins remain slightly dilated.

437. Friedreich's ataxia is characterized by:
    A. Pes cavus   B. Absent tendon reflexes
    C. Scoliosis   D. All of the above

438. All of the following are true about syndenham's chorea except:
    A. Common in adolescent girls
    B. A Late manifestation of rheumatic activity
    C. Long term prophylaxis, as for rheumatic fever, is essential
    D. None of the above

439. Frequency and urgency of urine is a feature of:
    A. Frontal lobe lesion   B. Parietal lobe lesion
    C. Spinal cord lesion   D. Medullary lesion

Ans.: 432. B   433. D   434. C   435. B   436. A   437. D   438. D
      439. B

440. **Periventricular calcification in a child with mental retardation is a feature of:**
   A. Tuberculoma
   B. Cysticercosis
   C. Toxoplasmosis
   D. Cytomegalovirus infection

441. **All of the following diseases may give rise to papilloedema except:**
   A. Vitamin A intoxication
   B. Polycythemia vera
   C. Severe anaemia
   D. Epilepsy

442. **A condition which may be aggravated or precipitated by a hot water bath:**
   A. Multiple sclerosis
   B. Parkinsonism
   C. Subacute sclerosing panencephalitus (SSPE)
   D. Epilepsy

443. **Infrequent causes of parkinsonism include all except:**
   A. Manganese poisoning
   B. Carbon monoxide poisoning
   C. Wilson's disease
   D. Obesity

444. **Drug which is said to produce parosmia:**
   A. Rifampicin
   B. Metronidazole
   C. Phenytoin
   D. Ethambutol

445. **Headache starting or worsening in the morning and associated with vomiting points towars:**
   A. Raised intracranial tension
   B. Migraine
   C. Temporal arteritis
   D. Tension headaches

446. **Ataxia telengiectasia is characterized by:**
   A. High IgA
   B. Low IgA
   C. High IgG
   D. Low IgM

447. **Myastheaia is often associated with each of the following except:**
   A. Hoshimoto disease
   B. Thyrotoxicosis
   C. SLE or pernicious anaemia
   D. None of the above

448. **Unsually powerfull muscles in early infancy but later slow relaxation of a muscle is characteristically seen in:**
   A. Myotonia congenital
   B. Myotonia atrophica
   C. Polymyositis
   D. Metabolic myopathies

449. **Transient Ischaemic attack in the carotid territory usually should not exceed:**
   A. 1 hour
   B. 6 hours
   C. 1 day
   D. 3 days

450. **Two point discrimination is best tested at:**
   A. Sole of leg
   B. Tip of tongue
   C. Finger tips
   D. Leg

Ans.: 440. C   441. D   442. D   443. D   444. D   445. D   446. D
     447. D   448. A   449. C   450. B

451. **Horner's syndrome is characterized by all of the following except:**
    - A. Ptosis
    - B. Miosis
    - C. Enophthalmos
    - D. Paralysis

452. **Total paralysis of oculomotor nerve causes the ipsilateral defects which include the following except:**
    - A. Ptosis
    - B. Dilation of pupil
    - C. Loss of accommodation reflex
    - D. Lateral gaze paralysis

453. **Agenesis of corpus collasum is best diagnosed by:**
    - A. Carotid angiography
    - B. EEG
    - C. Echoencephalogram
    - D. Pneumoencephalogram

454. **Microcephaly is frequently associated with:**
    - A. Postmaturity
    - B. Infants of diabetic mothers
    - C. Cytomegalic inclusion bodies
    - D. Extended breech delivery

455. **Disc prolapse is most commonly seen between:**
    - A. C_7C_8
    - B. $T_{12}L_1$
    - C. L_4L_5
    - D. L_5L_1

456. **"Lacunar skull" is most often associated with:**
    - A. Arnold Chiari syndrome
    - B. Sturge Weber syndrome
    - C. Meningoencephalocele
    - D. Temporal lobe tumors

457. **Headache and diplopia are the only presenting features of:**
    - A. Subacute sclerosing panencephalitis (SSPE)
    - B. Herpes simplex encephalitis
    - C. Encephatitis lethargica
    - D. Acute lymphocytic choriomeningitis

458. **Parkinsonism may be caused by all except:**
    - A. Reserpine
    - B. Alpha methyldopa
    - C. Metoclopramide
    - D. Phenytoin sodium

459. **"Barber's chair" sign may occur in all of the following except:**
    - A. Spinal cord compression
    - B. Syringomyelia
    - C. Multiple sclerosis
    - D. Vitamin A intoxication

460. **Phenobarbitone should not be given in combination with:**
    - A. Carbamazepine
    - B. Valproic acid
    - C. Primidone
    - D. Acetazolamide

Ans.: 451. D   452. D   453. D   454. C   455. D   456. C   457. C
458. D   459. D   460. C

461. **The most common reason of epileptie fit while stopping antiepileptic drugs after an adequate and effective period of treatment is:**
   A. Progressive neurological disease
   B. Tolerance to antiepileptic drugs
   C. Precipitants of epilepsy like fever, stress etc.
   D. Sudden withdrawal

462. **All may be the complications of status epilepticus except:**
   A. Metabolic disturbances   B. Psychosis
   C. Brain damage   D. Death

463. **The most common cause of status epilepticus in patients with a prior seizure disorder is:**
   A. Uncontrolled or resistant epilepsy
   B. Abrupt cessation of anti-convulsion
   C. Metabolic disorders
   D. Progressive neurological disease

464. **To avoid risk of respiratory depression, the drug of choice in the control of status epileplicus which can be given in a plastic syringe is:**
   A. Diazepan   B. Diphenylhydantoin
   C. Phenobarbitone   D. Paraldehyde

465. **Which of the following symptoms parkinsonism first responds to levodopa :**
   A. Tremors   B. Rigidity
   C. Akinesia   D. Gait

466. **Which of the following type of Parkinsonism does not respond well to levodopa:**
   A. Idiopathic   B. Arteriosclerotic
   C. Post head injury   D. Postencephalitic

467. **Transient, rapid deterioration in symptoms (akinesia, tremors and rigidity) developing in minutes and clearing up spontaneously in a Parkinsonian patient taking levodopa indicates:**
   A. Tolerance to drug   B. Toxic levels
   C. Drug induced Parkinsonism   D. "On off" phenomenon

468. **To avoid gastrointestinal side effects such as nausea, vomiting and anorexia with levodopa therapy, best is:**
   A. Take drug with food   B. Give antacids
   C. Metoclopramide   D. Trifluopromazine

469. **All of the following drugs should not be given with levodopa except:**
   A. Reserpine   B. Propranolol
   C. Penfluridol   D. Pyridoxine

Ans.: 461. D   462. B   463. B   464. B   465. C   466. D   467. D
      468. A   469. B

470. **CSF pressure may be increased in all of the following except:**
- A. Subarachnoid hemorrhage
- B. Meningism
- C. Benign intracranial hypertension
- D. Spinal tumor

471. **Traumatic tap differs from subarachnoid hemorrhage by all of the following except:**
- A. Absence of xanthochromia
- B. Pressure usually normal
- C. Absence of clotting
- D. Blood staining differes from tube to tube

472. **Lower motor neuron type of facial palsy is characterized by:**
- A. Upper part of face unaffected
- B. No reaction of degeneration
- C. Normal nerve conduction and FMG
- D. Loss of emotional movement of facial muscles

473. **The presence of weakness or abolition of emotional movements of face with retention of voluntary movements (Mimic paralysis) is seen in lesions of:**
- A. Upper pons
- B. Lower pons
- C. Anterior part of frontal lobe
- D. Posterior part of frontal lobe

474. **Which of the following is known as Bell's phenomenon:**
- A. Loss or wrinkling of forehead and also frowning
- B. Involuntary blinking abolished
- C. On attempting closure of eyes, eyeball turns upwards and outwards
- D. On showing the teeth, the lips do not separate on affected side

475. **Which of the following is the least common cause of bilateral lower motor neurone facial palsy:**
- A. Acute infective polyneuritis
- B. Leprosy
- C. Post-diphtheric
- D. Bell's palsy

476 **Which of the following muscles may be involved in Bell's paralysis:**
- A. Stylohyoid
- B. Cricothyroid
- C. Anterior belly of digastaic
- D. Sternohyoid

477. **All of the following are true about Tie Douloureux except**
- A. Common after middle age
- B. More common in men
- C. Sensory division of fifth nerve involved
- D. Unilateral, intermittent lancinating pains in the face precipitated by chewing, eating, drinking, talking or the face

---

Ans.: 470. D   471. C   472. D   473. C   474. C   475. D   476. A
477. B

478. **The Commonest clinical type of Landry-Gullain Barre syndrome is:**
    A. Mononeuritic or Polyneuritis
    B. Cerebral type
    C. Myelitie
    D. Bulbar

479. **All of the following drugs may cause polyneuritis except:**
    A. Vincristine   B. Nitrofuratoin
    C. INH   D. Levodopa

480. **The deficiency of all of the following vitamins may cause polyneuropathy except:**
    A. Thiamine   B. Riboflavin
    C. Pyridoxine   D. B_{12}

481. **Carcinomations myasthenia differs from true myasthenia by all of the following except:**
    A. Onset in later age
    B. Ocular muscles more involved than limp muscles
    C. Tendon reflexes usually diminished
    D. Power improves following exercise but poorly with neostigmine

482. **All are genetically determined myopathies except :**
    A. Polymyositis
    B. Malignant hyperpyrexia syndrone
    C. Hyperkalemic type of periodic paralysis
    D. Normokalemic type of periodic paralysis

483. **The treatment of choice in Bell's palsy is :**
    A. Corticosteroids   B. Physiotherapy
    C. Carbamazepine   D. Erythematous rash

484. **The commonest side effect of carbamazepine is:**
    A. Dizziness   B. Nausea
    C. Diplopia   D. Erythematous rash

485. **In sciatica, the straight leg raising test of lasegue is useful in determining all of the following except:**
    A. Diagnosis   B. Severity
    C. Progress of illness   D. Respose to improvement

486. **Which of the following nerve roots are most often involved in thoracic inlet syndrome:**
    A. C_6C_7 B. C_7C_8
    C. C_8T_1 D. T_1T_2

Ans.: 478. C   479. D   480. B   481. B   482. A   483. B   484. A
         485. C   486. C

487. All of the following are features of acute infective polyneuritis (Landry-Gullain-Barre syndrome) except:
A. Quadriplegia
B. Involvement of cranial nerves II, III, VI and VIII
C. Loss of superficial and deep reflexes
D. Urinary incontinence

488. Which of the following is not a autosomal dominant muscular dystrophy:
A. Facioscapulohumeral dystrophy
B. Limb-Girdle muscular dystrophy
C. Distal muscular dystrophy
D. Oculopharyngeal muscular dystrophy

489. All of the following are features of myotonia atrophica except:
A. Cataract
B. Frontal baldness in males
C. Mental retardation or dementia
D. Spontaneous remission not uncommon

490. Which of the following is not true about Pseudo-hypertrophy muscular dystrophy (Duchenne type)
A. Common in male children
B. Autosomal dominant inheritance
C. Gower's sing is pathognomic
D. Knee jerks disappear before ankle jerks

491. Which of the following is a foramen magnum sign in cranio vertebral anomalies:
A. On closing one hand, the other also closes (Mirror movements)
B. Wasting of upper limbs due to infarction
C. Loss of position sense at distal interphalangeal joint
D. A component of field's triad (Shortneck, low hair line and restriction of neck movements)

492. All are components of Wallenburg's syndrome (lateral medullary syndrome) except:
A. Dysphagia
B. Burning sensation on same side of face with loss of pain and thermal sensibility on the opposite site of body below face
C. Ataxia and intention tremors on limbs on opposite side with hypotonia
D. Horner's syndrome

Ans.: 487. D   488. B   489. D   490. B   491. C   492. C

493. Match the following:

I.   Anterior cerebellum   (i)   Equilibrium
     (Paleocerebellum)

II.   Posterior cerebellum   (ii)   Posture and muscle tone
     (Neocerebellum)

III.   Flocculonodular   (iii)   Coordination of skilied movements
     (Archicerebellum)

A. I (i)   II (ii)   III (iii)
B. I (ii)   II (iii)   III (i)
C. I (iii)   II (ii)   III (i)
D. I (iii)   II (i)   III (ii)

494. Disturbance in the coordination of various muscles and muscle groups participating in a movement e.g. extending the trunk backwards without simultaneous flexion of the knees, thus loosing balance, is seen on a cerebellar lesion and is called:

A. Barany's test   B. Dysmetria
C. Dyssynergy   D. Dysdiadochokinesia

495. All of the following intracranial structures are sensitive to pain except:

A. Internal carotid artery   B. Sagittal venous sinus
C. Corpus collasum   D. Dura at the base of brain

496. Which of the following is the commonest type of migraine:

A. Classic migraine   B. Common migraine
C. Complicated- migraine   D. Ophthalmic migraine

497. Which of the following symptoms are most often seen in aura of classic migraine:

A. Headache   B. Nausea
C. Dizziness   D. Visual symptoms

498. Common migraine is characterized by all except:

A. Characteristic aura
B. Longer duration of headache than classic type
C. Sensitivity to light and noise
D. Often occurs on weakends and holidays

499. The area of brain most frequently affected (complicated) by migraine is:

A. Parietal lobe   B. Occipital lobe
C. Temporal lobe   D. Brain stem

500. All of the following factors may provoke migraine except :

A. Tyramine containing foods   B. Caffeine
C. Oral contraceptives   D. Propranolol

Ans.: 493. B   494. C   495. C   496. B   497. D   498. A   499. B
      500. D

501. Which of the following drugs is not ofen useful in the prophylaxis of migraine:
  A. Verapamil   B. Pizotifen
  C. Amitriptyline   D. Thiazide

502. The electrical phenomenon that occurs in single neurons that generates each interictcal, epileptiform EEG spike is called:
  A. Forced normalization
  B. Paroxysmal depolarization shift (PDS)
  C. Triphasic wave pattern
  D. None of the above

503. Picture of an "Inverted champagne bottle" limb (thin legs contrast strikingly with the normal thigh) is characteristically seen in:
  A. Acute infective polyneuritis
  B. Polymyositis
  C. Peroneal muscular atrophy
  D. Amyotrophic lateral sclerosis

504. Match the following:

Motor neurone disease (clinical syndromes)	*Earliest presenting Symptom*
I. Progressive bulbar paralysis	(i) Weakness in muscle of one hand or forearm
II. Amytrophic lateral sclerosis	(ii) Dysphagia
III. Progressive musclar atrophy	(iii) Difficulty in walking

  A. I (i)   II (ii)   III (iii)
  B. I (ii)   II (iii)   III (i)
  C. I (ii)   II (i)   III (iii)
  D. I (iii)   II (ii)   III (i)

505. Which of the following reflexes disappear in the absence of functional connections between the spinal cord and the brain:
  A. Sweating reflex   B. Withdrawal reflex
  C. Swallowing reflex   D. Erection of penis

506. Two most useful test in a suspected case of spinal cord compression is:
  A. Barber's chair sign   B. Myelography
  C. Queckedstedt's test   D. Sensory evoked potential

507. Neural tube defect is a teratogenic effect of:
  A. Diphenylhydantion   B. Ethosuximide
  C. Carbamazepine   D. Valproic acid

Ans.: 501. D   502. B   503. C   504. B   505. C   506. C   507. D

508. Diabetic neuropathy is associated with all except:
- A. Asymmetric mononeuropathy multiplex
- B. Acute mononeuropathy
- C. Sensory polyneuropathy
- D. None of the above

509. All of the following are associated with von Reckling-hausen's neurinomas
- A. Acoustic neurinomas
- B. Optic glioma
- C. Pheochromocytoma
- D. Diabetic insipidus

510. The structure in the brain which is most susceptible to hyperthermia is:
- A. Cranical nerve nuclei
- B. Putamen
- C. Posterior hypothalamurs
- D. Purkinje cells of cerebellum

511. Heat stroke is characterized by all of the following except:
- A. Increase in serum potassium
- B. Increase in blood urea nitrogen
- C. Respiratory alkalosis
- D. Hypocalcemia aad hypophosphatemia

512. Neuroleptie malignant sysdrome (characterized by muscular rigidity, hyperthermia, autonomic dysfunction) shows best response to:
- A. Penfluridol
- B. Alprazolam
- C. Dantrolene sodium
- D. Amoxapine

513. Hyperthemia due to central causes is characterized by all of the following except:
- A. Absence of sweating
- B. Loss of diurnal variation
- C. Resistance to cooling
- D. Resistance to antipyretics

514. Pyramidal signs with absence of ankle jerks may be seen in :
- A. Hypothyroidism
- B. Ribflavin deficiency
- C. Lathyrism
- D. Ankolysing spondylosis

515. Treatment of choice in acute migraine attacks is:
- A. Ergotamine tartarate
- B. Caffeine
- C. Propranolol
- D. Methysergide

516. Which of the following spinal cord segment is first affected in syringomyelia:
- A. C_7
- B. C_8
- C. T_1
- D. T_2

Ans.: 508. D   509. D   510. D   511. A   512. C   513. C   514. C
515. A   516. C

517. Which of the following is the most common diagnostic test for Wilson's disease:
   A. Serum copper below 90 mcg%
   B. Serum cereuloplasmin below 20 mcg%
   C. Urinary copper excretion more than 200 mcg/day
   D. Liver content of copper in excess of 250 mcg/gm dry weight

518. Which is the following agents is useful in the treatment of Wilson's disease:
   A. Penicillamine   B. EDTA
   C. BAL   D. All of the above

519. In Huntington's chorea, lesion lies in:
   A. Anterior horn cells
   B. Corpus striatum
   C. Motor cranial nuclei in brain stem
   D. Hindbrain

520. Burning feet syndrome (intense discomfort in feet in absence of objective sensory or motor signs) is seen in deficiency of:
   A. Thiamine   B. Pyridoxine
   C. Pantothenic acid   D. Cyanacobalamin

521. In Wernicke's encephalopathy, petechial hemorrhages occur in:
   A. Frontal lobe   B. Hippocampus
   C. Mamiltary bodies   D. Corpus collasum

522. Following are the precipitating factors of multiple sclerosis except:
   A. Emotional disturbance or fatigue
   B. Pregnancy
   C. Carbon monoxide poisoning
   D. Baclofen

523. All of the following CSF changes strongly support the diagnosis of multiple sclerosis except:
   A. Colloidal gold curve paretic or liuetic
   B. Gamma globulin content of total spinal fluid protein more than 14%
   C. Presence of oligoclonal bands in CSF
   D. Albumino – cellular dissociation

524. All are the components of Charcot's triad, present in late stage of multiple sclerosis except:
   A. Ataxia   B. Nystagmus
   C. Intention tremor   D. Scanning speech

Ans.: 517. D   518. D   519. B   520. C   521. C   522. D   523. D
      524. A

525. **All are the criteria of diagnosis of multiple sclerosis except:**
    A. Age of onset 10-15 years
    B. Lesions dissociated in time and place with predominant affection·
       of white matter
    C. Disease lasting for more than one month
    D. All other causes have been ruled out

526. **Long term survival is possible after surgical removal of:**
    A. Astrocytoma   B. Medullobilastoma
    C. Acoustic neuroma   D. Oligodendroglioma

527. **All of the following may be accompanied by raised intracranial pressure except:**
    A. Personality changes   B. Generalized epilepsy
    C. Early morning headache   D. Taachycardia

528. **"Gaze nystagmus" may be seen in all except:**
    A. Wernicke"s encephalopathy B. Brain stem glioma
    C. Parkinsonism   D. Multiple sclerosis

529. **Earliest manifestation of papilloedema is:**
    A. Engorgement of retinal veins
    B. Oblileration of physiological cup
    C. Pink discoloration of optic disc
    D. Blurring of disc margin on nasal side

530. **Thalamic sysdrom (diffuse burning pain on the opposite side of the body) is seen in occlusion of :**
    A. Anterior cerebral artery   B. Middle cerebral artery
    C. Posterior cerebral artery   D. Anterior communicating artery

531. **A disorder of peripheral nerves and the optic nerves and rarely muscle weakness:**
    A. Subacute combined dogeneration
    B. Pellagra
    C. Nutritional polyneuropathy
    D. Strachan's sysdrome

532. **An absent plantar response is seen in all except:**
    A. Cold feet   B. Spinal shock
    C. Sensory loss in S_2 distribution D. Sleep

533. **Papilloedema may occur in:**
    A. Cushing 's syndrome
    B. Addison's disease
    C. Hypoparathyroidism
    D. Hypercapnia without lung disease

Ans.: 525. C   526. C   527. D   528. C   529. A   530. C   531. D
       532. D   533. D

534. Migraine may be precipitated by:
A. Thiazide
B. Propranolol
C. Reserpine
D. Caffeine

535. All are rich sources of tyramine (a precipitant of migraine) except:
A. Red wine
B. Cheese
C. Mutton
D. Sherry

536. True about visual symptoms of migraine include all except:
A. Photophobia
B. Due to dilatation of cortical arteries
C. Seldom last longer than half an hour
D. Include temporary defects in the visual field

537. Upper motor neurone lesion is caused by:
A. Wilson's disease
B. Fascio – scapulo-humeral muscular dystrophy
C. Cold exposure
D. Subacute combined degeneration

538. All of the following are useful in the treatment of multiple scierosis except:
A. Baclofen
B. Corticotrophin
C. Azathioprine
D. Vitamin B_{12}

539. Acute transverse myelitis is most common in which region of spinal cord:
A. Cervical
B. Dorsal
C. Lumbar
D. Sacrococcygeal

540. Priapism is common if acute transverse myetitis is seen in the region
A. Cervical
B. Thoracie
C. Lumbar
D. Sacrococcygeal

541. Which of the following presentation is common at the onset of acute myelitis:
A. Paraplegia in flexion
B. Paralegia in extension (stage of flaccidity)
C. Paraplegia in extension (stage of spasticity)
D. Mass reflex

542. All are common symptoms of intramedullary intradural tumors except:
A. Paraesthesia
B. Muscle fasciculation
C. Radicular pains
D. Dissociation of sensations

Ans.: 534. C   535. C   536. B   537. D   538. D   539. B   540. B
541. B   542. C

543. **Which of the following is a frequent symptom of extra medullary intradural tumor:**
   A. Muscle fasciculation
   B. Bladder and rectal involvement
   C. Muscle atrophy and trophic skin changes
   D. Spinal fluid changes

544. **Brown –Sequard syndrome is common in**
   A. Acute transverse myelitis
   B. Multiple sclerosis
   C. Extramedullary intradural tumor
   D. Intramedullary intradural tumor

545. **Match the following:**

Reflex		Nerve root	
I.	Biceps	(i)	$L_1 - L_2$
II.	Supinator	(ii)	$S_1 - S_2$
III.	Ankle jerk	(iii)	$C_5 - C_6$
IV.	Cremastric	(iv)	$C_6 - C_7$

546. **Conus medullaris lesion in contrast to Cauda equina is characterized by all of the following except:**
   A. Onset sudden and bilateral
   B. Saddle shaped anesthesia
   C. Knee and ankle jerks absent
   D. Bladder involvement early and marked

547. **All are the causes of combination of upper motor neurone paralysis in upper limps, Except:**
   A. Syringomyelia   B. Multiple sclerosis
   C. Friedreich's ataxia   D. Amyotrophic lateral sclerosis

548. **All are the clinical features of Pott's paraplegia except:**
   A. Paraparesis or paraplegia which may be symmetrical or not
   B. Root pains and paraesthesia
   C. Sphincter involvement early
   D. Spinal deformity (gibbus)

549. **Subacute combined degeneration in contrast to Tabes dorsalis has all the following signs except:**
   A. Mental changes
   B. Pupils normal
   C. Gross ataxia
   D. Evidence of peripheral nerve affection but CSF normal

Ans.: 543. D   544. C   545. C   546. C   547. C   548. C   549. C

550. "Triad of symptoms" in Tabes dorsalis includes all except:
   A. Argyil Robertson pupils
   B. Absent Tendon reflexes
   C. Charcot Joints
   D. Positive Romberg's sign

551. Hirsutism, a troublesome side effect in females, may be seen with :
   A. Diphenyhydantoin   B. Primidone
   C. Trimethadione   D. Valproic acid

552. Abadie's sign in Tabes dorsalis is :
   A. Patient walks on a wide base with eyes fixed to the ground
   B. Butterfly type of anesthesia on face
   C. Paroxysmal painful disorders of function of various viscera
   D. Anaesthesia of tendon achilis

553. Match the following:

Pupil		*Lesion*	
I.	Fixed dilated	(i)	Third nerve compression
II.	Constricted, non-reacting	(ii)	Mid-brain lesion
III.	Constricted but reacting	(iii)	Thalamic lesion
IV.	Unequal	(iv)	Medullary lesion

   A. I (iv)   II (i)   III (ii)   IV (iii)
   B. I (iv)   II (iii)   III (ii)   IV (i)
   C. I (i)   II (ii)   III (iii)   IV (iv)
   D. I (iv)   II (ii)   III (iii)   IV (i)

554. Which of the following is not a component of triad of pontine hemorrhage:
   A. Pyrexia   B. Pain
   C. Paraesthesia   D. Pin point pupil

555. Which of the following is not an early presentation of spinal subarachnoid hemorrhage of lumbago sciatica type:
   A. Pain in back and lower limb
   B. Flaccid paralysis or paresis of lower limbs
   C. Sphincter disturbance
   D. Kernlg's sign

556. Commonest site of berry aneurysm (subarachnoid hemorrhage is):
   A. Anterior cerebral artery
   B. Middle cerebral artery
   C. Basilar artery system
   D. Internal carotid artery intracerebrally

Ans.: 550. C   551. A   552. D   553. D   554. C   555. B   556. A

557. All are true about Wernicle enephalopathy except:
A. Nystagmus common
B. Loss of papillary reflex
C. Alcoholism is the only cause
D. Ocular movements may be impaired

558. Which of the flowing causes "inverse steal syndrome" shunting blood to ischemic zone:
A. Cyclendelate   B. Papaverine
C. Aminophylline   D. Frusemide

559. Match the following :

Pupil		*Lesion*	
I.	Millard-Gubber syndrome	(i)	Third nerve palsy on side of lesion with Tremors and ataxia on opposite side
II.	Foville's syndrome	(ii)	Third nerve palsy with crossed hemiplegia
III.	Weber's syndrome	(iii)	VI and VII nerves involved with crossed Hemiplegia
IV.	Benedict's sysdrome	(iv)	Nuclear type of facial palsy with crossed Hemiplegia

A. I (i)   II (ii)   III (iii)   IV (iv)
B. I (ii)   II (i)   III (iii)   IV (iv)
C. I (iv)   II (iii)   III (ii)   IV (i)
D. I (iii)   II (iv)   III (ii)   IV (i)

560. Dejerine's syndrome (medial medullary syndrome) is characterized by all except:
A. Ipsilateral flaccid tongue weakness
B. Contralateral hemiplegia
C. Ipsilateral anesthesia of face
D. Contralateral loss of position and vibration sense

561. Wallenberg's syndrome (lateral medullary syndrome) has except:
A. Abrupt onset with vertigo
B. Ipsilateral anaesthesia of face and contralateral anaesthesia of limbs and trunk
C. Contralateral intention tremor
D. Horner's sysdrome and nystagmus

562. Commonest site of cerebrovascular disease causing hemiplegia :
A. Cortex   B. Internal capsule
C. Midbrain   D. Pones

Ans.: 557. C   558. C   559. C   560. C   561. C   562. B

563. Cerebrovasular accident of hemorrhage type differs from thrombotic type by all of the following except:
   A. Hypertension almost invariable
   B. Convulsions and severe headache common
   C. Signs like stiff neck, conjugate deviation of eyes and Cheyne Strokes breathing common
   D. Premonitory symptoms are frequent

564. Which of the following cerebrovascular accident has highest mortality:
   A. Embolic
   B. Thrombotic
   C. Hemorrhage type
   D. Equal in all of the above

565. Ophthalmodynamometry indicates
   A. External carotid artery disease
   B. Internal carotid artery disease on low pressure side of ophthalmic artery
   C. Internal carotid artery on high pressure side
   D. Raised intracranial pressure

566. All may be the causes of hemiplegia of slow onset except:
   A. Cerebral tumor
   B. Cerebral abscess
   C. Chronic subdural hematoma
   D. Hysteria

567. Which of the following artery is most frequently affected in embolic or thrombotic cerebrovascular disease:
   A. Anterior cerebral
   B. Anterior communicating
   C. Middle cerebral
   D. Posterior cerebral

568. Which of the following is lease sedating and least respiratory depressant:
   A. Diazepam
   B. Phenobarbitone
   C. Phenytoin
   D. Paraldehyde

569. All are contraindications of Diphenylhydantoin except:
   A. Elderly
   B. Hepatic dysfunction
   C. Cardiac arrythmias
   D. Hypertension

570. Which of the following is a carbohydrate inhibitor useful in petit mal epilepsy:
   A. Aminoglutethimide
   B. Paramethadione
   C. Sulthiame
   D. None of the above

571. The disadvantage or limitation of EEG that it does not exclude:
   A. Organic brain disease
   B. Epilepsy
   C. Tumor (benign or malignant)
   D. All of the above

Ans.: 563. D   564. C   565. B   566. D   567. C   568. D   569. D
570. C   571. D

572. EEG is most useful in the diagnosis of:
A. Schisophrenia
B. Hysteria
C. Dementia
D. Inclusion body encephalitis

573. Following are contrairdications (misuses) of EEG except:
A. If physician advises treatment on the basis of tracing
B. If it is dependent on normal report
C. If patient serious or precariously injured
D. For investigating a comatosed patient whose physical signs are not of localizing nature.

574. The uses of an EEG is that it is used to:
A. Investigate a patient with attacks of disturbance in consciousness
B. Investigate unexplained intellectual deterioration
C. Investigate sudden unexplained episodes of disturbed behaviour
D. Rule out seizure disorder

575. Following are the normal calcifications seen in brain in X-ray skull or CT scan except:
A. Pineal gland
B. Choroid plexus
C. Basal ganglion
D. Falx

576. Match the following:

Calcification
I. Rounded balls (Brain stones)
II. Curvilinear
III. Albl's ring
IV. Tramline calcification

Cerebral lesion
(i) Sturge weber syndrome
(ii) Aneurysm near sella tursica
(iii) Tuberous sclerosis
(iv) Atheroma of carotid artery

A. I (ii)   II (iii)   III (iv)   IV (i)
B. I (iii)   II (iv)   III (ii)   IV (i)
C. I (iv)   II (iii)   III (ii)   IV (i)
D. I (i)   II (ii)   III (iii)   IV (iv)

577. Dysphonia may be the consequence of all except:
A. Supranuclear bulbar palsy
B. Cerebellar disease
C. Parkinsonism
D. Myasthenia gravis

578. Of the following structures, which is better visualized in a CT scan (head) in comparison to MRI (Magnetic Resonance Imaging):
A. Differentiation of Gray matter from white matter
B. Demyelinating diseases
C. Differentiation of infarct and tuberculomas
D. Better visualization of bone and small calcifications

Ans.: 572. D   573. D   574. D   575. C   576. B   577. B   578. D

579. The following are hard (reliable) signs of organic disease in the nervous system:
A . Frontal release signs (like snout reflex, sucking reflex etc.)
B. An extensor plantar response
C. Chvostek's sign
D. An absent 'gag' reflex

580. Dysphonic may be seen in all except:
A. A lesion in the speech area in dominant hemispheria
B. Supranuclear bulbar palsy
C. Myasthenia gravis
D. Parkinsonism

581. Which of the following is a symptom of myasthenia gravis:
A. Absent papillary reflexes   B. Muscular wasting
C. Ptosis   D. B + C

582. All are features of parkinsonism except:
A. Cogwheel rigidity   B. Pill rolling tremors
C. Festinant gait   D. Nystagmus

583. Impaired facial sensation is seen in all except:
A. Herpes zoster   B. Intrasranial aneurysm
C. Acoustic neuroma   D. Trigeminal neuralgia

584. All are the recognized features of an upper motor neuron seventh nerve palsy except:
A. Ptosis
B. Inability to wrinkle forehead
C. Weakness of masseter muscle
D. Preservation of emotional smiling

585. All of the following are reliable signs of upper motor neurone lesion except:
A. Babinski's sign
B. Positive Rossolimo's sign
C. Sustained ankle clonus
D. Generalized increase in tendon reflexes

586. All are true about progressive supranuclear palsy (Richardson steele-Oeszewski syndrome except:
A. Pseudobulbar palsy
B. Horizontal eye movements more restricted than vertical
C. Supranuclear ophthalmoplegia
D. Gait disturbance is a common presenting feature

Ans.: 579. B   580. A   581. C   582. D   583. D   584. D   585. D
586. B

587. **The earliest symptom of Hepatolenticular degeneration is:**
    A. Unintelligible speech   B. Dementia
    C. Tremor   D. Rigidity

588. **All are features of Benign intracranial hypertension (Pseudotumor cerebri):**
    A. Centricles of normal size
    B. Unaccompanied by an obstruction to CSF circulation
    C. Unexplained rise in CSF pressure
    D. All of the above.

589. **Subdural haematoma is most commonly seen in:**
    A. Infants'   B. Adolescents
    C. Middleaged   D. Elderly

590. **The treatment of choice in Binswanger's disease (Progressive subcortical encephalopathy) is:**
    A. Mannitoi   B. Nootropics like piracetam
    C. Antiepileptics   D. Antihypertensives

591. **All are the symptoms of the shy-drager syndrome (Progressive automic failure) except:**
    A. Postural hypotension   B. Dementia
    C. Sleep apnoea   D. Syncope

592. **Chronic intracranial abscess is usually characterized by all of the following except:**
    A. Papilloedema
    B. EEG abnormalities
    C. Fever
    D. Epilepsy in a high proportion even if successfully treated

593. **Very high CSF protein content is seen in all of the following except:**
    A. Acoustic neuroma   B. Guillain Barre syndrome
    C. Spinal neurofibroma   D. Neurosyphilis

594. **Vertical nystagmus is offen seen in :**
    A. Cerebellar lesions   B. Syringobulbia
    C. Congenital cataract   D. Drug intoxications

595. **Most common cause of reflex epilepsy is:**
    A. Buzzling noises   B. Loud music
    C. Monotonous sound   D. Flickering light

596. **All of the following antiepileptic drugs are nephrotoxic except:**
    A. Acetazolamide   B. Phensuximide
    C. Aminoglutethimide   D. Trimethadione

Ans.: 587. C   588. D   589. D   590. D   591. B   592. C   593. D
      594. D   595. D   596. C

597. ACTH is most useful as an antiepileptic in :
A. Recurrent myoclonic jerks in adults
B. Lennox Gestaut syndrome
C. Hypsarrhythmia
D. All of the above

598. Match the following:

Gait	Disease
I. "Frog" like	(i) Chorea
II. Marche a petits pas	(ii) Muscular dystrophy
III. Spastic Springing	(iii) Frontal lobe disease
IV. Jaunting	(iv) Lathyrism

A. I (iv) II (iii) III (ii) IV (i)
B. I (ii) II (iv) III (iii) IV (i)
C. I (i) II (ii) III (iii) IV (iv)
D. I (ii) II (iii) III (iv) IV (i)

599. Match the following:

Gait	Disease
I. Unilateral high steppage (Stapping type)	(i) Congenital cerebral diplegia
II. Festinant (Shuffling)	(ii) Alcoholic intoxication
III. Waddling (Ducklike)	(iii) Foot drop
IV. Reeling (Tottering)	(iv) Muscular dystrophy
V. Scissor	(v) Tabes dorsalies
VI. Stamping	(vi) Parkinsonism

A. I (i) II (iii) III (iv) IV (v) V (vi) VI (ii)
B. I (ii) II (iii) III (iv) IV (v) V (vi) VI (i)
C. I (iii) II (iv) III (v) IV (vi) V (i) VI (ii)
D. I (iii) II (vi) III (iv) IV (ii) V (i) VI (v)

600. Normal CSF level of glucose in mmol/L is:
A. 3.0   B. 8.0
C. 12.0   D. 30.0

601. All of the following are causes of increased gamma globulin levels in CSF except:
A. Multiple sclerosis
B. Fungal meningitis
C. Subacute sclerosing encephalitis
D. Guillain – Barre syndrome

602. On colloidal gold test (lange curve), normal CSF shows precipitation
A. First zone   B. Mid zone
C. End zone   D. None of the above

Ans.: 597. C   598. D   599. D   600. A   601. D   602. D

603. **Which of the following is common cause of pleocytosis, diminished chlorides and absent sugar in CSF:**
   A. Pyogenic meningitis   B. Tuberculous meningitis
   C. Viral meningitis   D. Fungal meningitis

604. **Which of the following, if present in CSF, indicates grave prognosis:**
   A. End zone on colloidal gold test
   B. Froin's sysdrome
   C. Absent CSF sugar
   D. Presence of acetone

605. **Normal pressure of CSF in mm of water is:**
   A. 1 – 4   B. 10 - 20
   C. 50 – 150   D. 150 – 220

606. **Which of the following cranial nerve is not involved in Tic douloreux:**
   A. V   B. VII
   C. VIII   D. IX

607. **Which of the following is best test in the diagnosis of cranial or temporal arteritis:**
   A. ESR   B. Gamma globulin levels
   C. Cerebral angiography   D. Temporal artery biopsy

608. **Which of the following is most useful investigation in the differential diagnosis of cerebral vascular lesion:**
   A. X-ray chest and skull   B. Pneumoencephalography
   C. Cerebral angiography   D. CT scan of head

609. **All of the following are causes of pleocytosis and increase of CSF protein without alteration in chloride and sugar levels:**
   A. Brain abscess   B. Aseptic meningitis
   C. Subarachnoid hemorrhage   D. Poliomyelitis

610. **Contraindications of pentoxyfylline include all except:**
   A. Severe haemorrhage (including retinal haemorrhage)
   B. Acute myocardial infarction
   C. Pregnancy
   D. Diabetes mellitus

611. **Which of the following drugs is most useful in Lennox Gestaut syndrome:**
   A. Phenobarbitone   B. Primidone
   C. Diphenythydantoin   D. Clonazepam

612. **Match the following:**
   Anticonvulsant   *Mechanism of action*
   I.   Phenobarbitone   (i)   GAB Amimetic
   II. Diphenythydantoin   (ii) Blocking posttetanic

Ans.: 603. A   604. D   605. C   606. C   607. D   608. C   609. D
   610. D   611. D   612. D

potentiation (PTP)

III. Succinimides   (iii) Inhibits excitatory and
                                   inhibitory activity within
                                   thalamic circuit

IV. Valproic acid   (iv) Stabilizing neuronal membrane
                                   potential

A. I (i)   II (iii)   III (iv) IV (ii)
B. I (iii)   II (iv)   III (ii) IV (i)
C. I (iv)   II (iv)   III (iii) IV (i)
D. I (iv)   II (ii)   III (iii) IV (i)

613. **Acetazolamide is an adjunet to other major anticonvulsants, especially when used in the management of seizure disorders aggravated by:**
A. Head injury   B. Fever
C. Major traquiffizers   D. Menses

614. **Alopecia may be a side effect of anticonvulsant:**
A. Ethosuximide   B. Diphenythydantoin
C. Carbamazepine   D. Valproic acid

615. **Hirsutism, a troublesome side effect in females, may be seen with:**
A. Diphenyihydantoin   B. Primidone
C. Trimethadione   D. Valproie acid

616. **Match the following:**

Drug   *Average dose (mg/kg body weight /day*

I.   Phenobarbitone   (i)   7 – 10
II.   Diphenylhydantoin   (ii)   1 – 5
III. Carbamazepine   (iii) 20-30
IV.   Valproic acid   (iv)   4 – 7

A. I (ii)   II (iv)   III (iii) IV (i)
B. I (ii)   II (iv)   III (i)   IV (iii)
C. I (iv)   II (ii)   III (i)   IV (iii)
D. I (ii)   II (i)   III (iv)   IV (iii)

617. **Match the following:**

Drug   *Therapeutic plasma range (µg/ml)*

I.   Phenobarbitone   (i)   4 – 10
II.   Carbamazepine   (ii)   40 – 150
III. Diphenythydantoin   (iii) 20 – 40
IV.   Ethosuximide   (iv)   10 – 20

Ans.: 613. D   614. D   615. A   616. B   617. C

618. **Which of the following is commonest cause of therapeutic failure with antiepileptic:**
    A. Inadequate doses
    B. Improper drugs or combinations
    C. Progressive neurological disease
    D. Non-compliance due to psychological factors or side effects of drugs.

619. **Radiculomyelopathy may be seen in:**
    A. Vitamin A intoxication
    B. Hypoparathyroidism
    C. Pellagra
    D. Flurosis

620. **Pseudoparalysis in an infant is suggestive of:**
    A. Acute rheumatic fever
    B. Vitamin D deficiency
    C. Vitamin C deficiency
    D. Marasmus

621. **The parasympathetic nerve supply to the urinary bladder is derived from the following segments of spinal cord:**
    A. L_2 and L_3
    B. S_2 and S_3
    C. T_{12} and T_1
    D. L_3 and L_4

622. **The following may be useful in the management of CSF rhinorrhoea except:**
    A. Repeated lumbar puncture
    B. Antibiotics
    C. Craniotomy and dural repair
    D. Nasal packing

623. **Acute extradural haematoma may have the following features except:**
    A. Tear of superior cerebral vein
    B. Fracture of temporal bone
    C. Presence of lucid interval
    D. Haematoma under temporalis muscle

624. **Commonest site of Prolapsed intervertebral disc is:**
    A. C5 – C6
    B. C7 – T1
    C. L4 – L5
    D. All sites have equal chances

625. **The only method to differentiate presenile dementias of Izheimer's and Pick's disease is by:**
    A. Immunoglodulin profile
    B. EEG
    C. CT Scan of head
    D. Brain biopsy

626. **Which of the following factors play an important role in the early manifestations of dementia:**
    A. Adequacy of personal coping mechanism
    B. Environmental stresses
    C. Family support
    D. All of the above

Ans.: 618. D   619. D   620. C   621. B   622. A   623. A   624. C
625. D   626. D

627. **Which of the following is an uncommon feature of early stage of dementia of Alzheimer type:**
   A. Problems with :word finding" and halting speech
   B. Denial of defects and rationalizations
   C. Decreased concentration for tasks, watching TV or reading
   D. Disorientation for time, place and person

628. **Pseudodementia is characterized by all of the following except:**
   A. Patients highlight their failures
   B. Pressure of "don't know" rather than approximate or wrong answers
   C. Attention, concentration and orientation are often intact
   D. Severe loss of recent memory only

629. **Which of the following is not true about Parkinsonism:**
   A. Rhythmic, pill rolling tremors
   B. Mask-like facies with reduced blinking
   C. Leadpipe or cog wheel type of rigidity
   D. Dementia-like syndrome is an early sign

630. **Temporal lobe lesion is characterized by all except:**
   A. Cortical deafness
   B. Auditory agnosias
   C. Kluver-Bucy syndrome in bilateral lesions
   D. Gerstmann syndrome

631. **Gerstmann syndrome, seen in dominant parietal lobe lesion, is characterized by all except:**
   A. Agraphia   B. Dyscalculia
   C. Agnosia (finger)   D. Anosognosia

632. **Which of the following is uncommon symptom in parietal lobe lesion:**
   A. Anosognosia
   B. Constructional apraxia
   C. Right-left disorientation
   D. Auditory agnosia or cortical deafness

633. **Best method to diagnose epilepsy is:**
   A. A reliable history   B. X-ray skull
   C. Electroencephalography   D. CT scan of head

634. **Coma cannot be presumed to be equivalent to deep sleep because:**
   A. In coma, although the EEG shows diffuse slowing, there are not the phasic shifts from NREM to REM sleep
   B. In come, there are no associated shifts in arousability as found in deep sleep
   C. Coma is characterized by distinct physiological alterations never found in normal sleep
   D. All of the above

Ans.: 627. D   628. D   629. D   630. D   631. D   632. D   633. A
   634. D

635. **"Punch drunk syndrome" may be characterized by all except:**
    A. Dementia
    B. Seizures
    C. Delirium
    D. None of the above

636. **Which of the following is most pathognomic sign in differentiating structural from non-structural causes of coma:**
    A. Increase or decrease is respiratory rate
    B. Presence or absence of light reflex
    C. Caloric stimulation of the auditory canal
    D. Motor responses like seizures, rigidity or release signs

637. **Post head trauma (PHT) syndrome may be characterized by all except:**
    A. Heightened startle response
    B. Throbbing headache, altered by postural changes and exertion
    C. Liability of mood
    D. Always presence of strong evidence of cognitive deficits on objective testing

638. **Korsakoff type of amnestic syndrome is often characterized by defect in:**
    A. Immediate memory
    B. Recent memory
    C. Remote memory
    D. All of the above

639. **The most common cause of senile and presenile dementia is:**
    A. Alzheimer's disease
    B. Pick's disease
    C. Creutz-Jacob disease
    D. Huntington's chorea

640. **The senile and presenile dementias are separated only by the arbitrary age of onset at or before:**
    A. 50 years
    B. 55 years
    C. 60 years
    D. 65 years

641. **The minimum score on the Glasgow coma scale is:**
    A. 0
    B. 1
    C. 3
    D. 5

642. **The commonest cause of epilepsy in geriatric age group is:**
    A. CNS infections
    B. Tumor
    C. Trauma
    D. Vascular

643. **Steroid therapy in tuberculosis meningitis helps to achieve:**
    A. Reduction of cerebral tension
    B. Reduction of cerebral edema
    C. Prevention of development of obstruction
    D. All of the above

Ans.: 635. D   636. B   637. D   638. B   639. A   640. D   641. C
642. D   643. D

644. Characteristics of congenital hydrocephalus include all the following except:
   A. Large head with wide and bulging fontanelle
   B. Transillumination
   C. Crackpot sign
   D. Convulsions

645. The structure in the brain most responsive to cold is:
   A. Anterior hypothalamus   B. Posterior hypothalamus
   C. Suprachiasmatic nuclei   D. Caudate nuclei

646. The best guide to the activity of polymyositis is :
   A. EMG   B. Muscle biopsy
   C. Serum ckenzyme level   D. Clinical improvement in power

647. In a traumatic lumbar puncture, which of the following RBC: WBC ratio will indicate a positive tap:
   A. 500 : 1   B. 1000 : 1
   C. 1500 : 1   D. 2000 : 1

648. Maximum daily therapeutic subdural tap (each side) should not exceed:
   A. 5 ml   B. 10 ml
   C. 20 ml   D. 50 ml

649. One of the leading etiopathogen for pyogenic meningitis in children in developing countries is:
   A. Esch. Coli   B. Meningococcus
   C. H. influenzae   D. Staphylococcus

650. Pseudotumor cerebri has been found in association with all of the following except:
   A. Hypo or hypervitaminosis A B. Sudden withdrawal of steroids
   C. Tetracycline toxicity   D. None of the above

651. Long term sequelae of acute infantile hemiplegia include all except:
   A. Seizures   B. Mental retardation
   C. Hemiparesis   D. Diabetes insipidus

652. When the speech is nonfluent, dysarthric, laborious, agrammatical and telegraphic, the lesion is in:
   A. Posterisuperior part of temporal lobe
   B. Precentral gyrus of frontal lobe
   C. Parietal lobe
   D. Anterior part of occipital lobe

653. Sensory inattention or neglect to bilateral simultaneous stimulation for the side contralateral to lesion is pathognomic of disorders of :
   A. Frontal lobe   B. Parietal lobe
   C. Temporal lobe   D. Occipital lobe

Ans.: 644. D   645. B   646. D   647. A   648. C   649. C   650. D
       651. D   652. B   653. B

654. Unawareness of illness or defect in a patient also with sensory inattention is known as:
- A. Astereognosia
- B. Atopognosia
- C. Anosognosia
- D. Astatognosia

655. Grestmann syndrome characterized by finger agnosia, dyscalculalia, agraphia and Right-left disorientation is seen in lesion to lobe:
- A. Dominant temporal
- B. Non–dominant temporal
- C. Dominant parietal
- D. Non-dominant parietal

656. All are functions lateralized to dominant hemisphere except:
- A. Verbal and ideational
- B. Analytic
- C. Geometric
- D. Sequential and digital

657. When a patient with cognitive deficits (especially in the deminant hemisphere lesions) is forced to face the situations where his deficits will be revealed, he may develop lack of cooperation, will be emotionally upset and may leave the interview. This is called:
- A. Alarm reaction
- B. Catastrophic reaction
- C. Dissociative hysteria
- D. Episodic decontrol syndrome

658. "Akinetic mutism" can be differentiated from "Locked in syndrome" by all of the following except:
- A. Lesion in the frontal lobe
- B. No impairment of cranial nerves
- C. Absence of any paralysis
- D. Absence of urinary or faecal incontinence

659. Release of primitive reflexes or signs such as grasp, palmomental, snout and sucking reflect on injury to:
- A. Frontal lobe
- B. Parietal lobe
- C. Temporal lobe
- D. Occipital lobe

660. Match the following:
- I.   Transversegyri of Heschl   (i)   Taste
- II.   Prepiriform cortex of frontal   (ii) Motor activity
         and temporal lobes
- III. Inferior most part of postcentral (iii) Small
         gyrus
- IV.   Brodmann areas 4, 6, 8   (iv) Hearing
- A. I (ii)   II (iv)   III (iii)   IV (i)
- B. I (iv)   II (i)   III (iii)   IV (ii)
- C. I (iii)   II (iii)   III (i)   IV (ii)
- D. I (iii)   II (iv)   III (i)   IV (ii)

Ans.: 654. C   655. C   656. C   657. B   658. D   659. A   660. C

661. **Which of the following is often the site of origin of Jacksonian epilepsy :**
   A. Orbitofrontal area
   B. Frontal lobe
   C. Prerplandic gyrus
   D. Postrolandic gyrus

662. **Which of the following is most probable case of epilepsy in early adulthood:**
   A. Tumor
   B. Vascular disease
   C. Congenital defects
   D. Trauma

663. **Which of the following is the best diagnostic aid in differentiating epileptic fit from hysterical type:**
   A. A reliable history
   B. EEG with sphenoidal leads
   C. Magnetic Resonance Imaging (MRI)
   D. Abreaction with intravenous pentothal

664. **Which of the following is the most frequent cause of fits in a made adult well maintained on antiepileptic drugs:**
   A. Infrequent or irregular intake of drugs
   B. Precipitants of epilepsy like sleep, hunger etc
   C. Toxic serum levels of antiepileptic drugs
   D. Pseudoseizures

665. **Which of the following are of brain is believed to be the site of seizure discharges causing a syndrome known as "Transient Global Amnesia":**
   A. Frontal lobe
   B. Parietal lobe
   C. Hippocampus
   D. Hypothalamus

666. **Sudden attacks of generalized or focal loss of postural tone sparing consciousness which may be initiated by adrupt emotional changes such as laughter, anger, fear or surprise is known as:**
   A. Sleep attacks
   B. Epileptic fugue
   C. Catalepsy
   D. Cataplexy

667. **Which of the following group of drugs considered to be the treatment of choice for narcolepsy:**
   A. Stimulants like dextromphetamine
   B. Nonsedating antidepressants
   C. Benzodiazepines such as clonazepam
   D. Antiepileptics like valproate

668. **Match the following:**

Antionvulsant		*Effect*	
I.	Phenylside chain	(i)	Clonic seizure and petit mal activity
II.	Alkyl side chains	(ii)	Tonic seizure

Ans.: 661. C   662. D   663. A   664. A   665. C   666. D   667. B
668. D

III.   Carbon side chains   (iii) Sedative effects

A. I (ii)   II (iii)   III (i)
B. I (i)   II (iii)   III (ii)
C. I (iii)   II (i)   III (ii)
D. I (ii)   II (i)   III (iii)

669. **The alpha rhythm of EEG indicates all of the following except:**
   A. Best seen when subjects eyes are closed
   B. Indicate that the subject is awake
   C. Has amplitude about 50 microvolts, frequency 8 – 12 Hz
   D. Most marked in the frontal area

670. **Alpha blaok (Replacement of alpha rhythm by fast, irregular low voltage activity with nondominant frequency) is seen in:**
   A. When eyes are open
   B. Sensory stimulation
   C. Solving problems with concentration
   D. All of the above

671. **Sleep spindles resemble the frequency of :**
   A. Delta waves   B. Theta waves
   C. Alpha waves (Berger waves) D. Beta waves

672. **The frequency of alpha rhythm is decreased by all of the following except:**
   A. Low blood glucose level
   B. A low level of adrenal glucocorticoid hormones
   C. A high arterial partial pressure of CO_2
   D. A high body temperature

673. **A child demonstrates irregular, spasmodie, involuntary movements of the limbs and facial muscles, is most likely to have a lesion in:**
   A. Pre central gyrus of cortex   B. Post central gyrus of cortex
   C. Midline cerebellum   D. Caudate nucleus

674. **Most common age group of febrile convulsions is:**
   A. 3 months to 6 years   B. 9 months to 2 years
   C. 1 to 7 years   D. 4 to 12 years

675. **"Lupus like syndrome" may be a side effect of:**
   A. Phenobarbitone   B. Diphnylhydantion
   C. Ethosuximide   D. Valproic acid

676. **Febrile convulsions are most often secondary to infections of :**
   A. CNS   B. Gastrointestinal tract
   C. Respiratory tract   D. Urinary tract

Ans.: 669. D   670. D   671. C   672. D   673. D   674. B   675. B
   676. C

677. **Predominant neck muscles weakness is seen in all of the following except:**
    A. Duchenne's muscular dystrophy
    B. Myotonia dystrophia
    C. Poliomyositis
    D. Myasthenia gravis

678. **Headache may be produced by all of the following except:**
    A. Dilation of intra cranial blood vessels
    B. Pressure of blood in CSF
    C. Mechanical damage to parietal lobe
    D. Loss of CSF following lumbar puncture

679. **Which of the following side effect of levodopa warrant its discontinuation:**
    A. Nausea and vomiting
    B. Potural hypotension
    C. Choreiform movements
    D. Sinus tachycardia

680. **Which of the following side effects of Levodopa is prevented by decarboxylase inhibitors:**
    A. Postural hypotension
    B. Abnormal involuntary movements
    C. Psychiatric disturbances
    D. Nausea and vomiting

681. **Which of the following is not a side effect of amantadine:**
    A. Epilepsy
    B. Mania
    C. Restlessness
    D. Postural hypotension

682. **All of the following are associated with peripheral neuritis except:**
    A. Diabetes mellitus
    B. Infectious mononucleosis
    C. Boeck's sarcoidosis
    D. Amyotrophic lateral sclerosis

683. **All of the may induce seizures or increase seizure tendency except:**
    A. Chlorpromazine
    B. Oral contraceptives
    C. Imipramine
    D. Acetazolamine

684. **All are sings of lower motor neuron disease except:**
    A. Spasticity
    B. Individual muscles may be affected
    C. Fasciculation
    D. Marked muscle atrophy

685. **All are signs of uppere motor neuron lesion exept:**
    A. Spasticity
    B. Fascicular twitches
    C. Normal reactions to galvanic and faradic current
    D. Babinski"s sign

Ans.: 677. A   678. C   679. C   680. D   681. D   682. D   683. D
684. A   685. B

686. All are the causes of "Albumino cytologic dissociation' in CSF except:
   A. Spinal cord obstruction
   B. Polyradicultis (Gullain-Barre syndrome)
   C. Poliomyelitis
   D. Syphilitic meningitis

687. A"Cobweb" type of fibrin clot in CSF is characteristic of :
   A. Poliomyelitis   B. Polyradiculitis
   C. Syphilitic meningitis   D. Tuberculous meningitis

688. All of the following are caused of decreased tension of cerebrospinal fluid except:
   A. Subdural hematoma   B. Uremia
   C. Spinal subarachnoid block   D. Repeated lumbar punctures

689. Which of the following is the treatment of the choice in brain abscess:
   A. Antibiotics   B. Steroids
   C. A + B   D. Surgery

690. Guillain-Barre syndrome may occur as a part of which of the following disease:
   A. Hodgkin's disease   B. Burkitt's lymphoma
   C. Infectious mononucleosis   D. Neurosyphilis

691. Which of the following is most serious complication of meningovascular syphilis:
   A. Epilepsy   B. Meningitis
   C. Vertigo   D. Hemiplegia

692. Which of the following type of neurosyphilis is often characterized by "shooting pains", "Charcot Joint" and "Gastric and rectal crises":
   A. Meningovascular syphilis   B. Tabes dorsalis
   C. General paresis   D. Meningitis

693. All of the following are early symptoms of multiple sclerosis noted frequently except:
   A. Diplopia
   B. Disturbed motor function
   C. Disturbed urinary bladder control
   D. Cerebellar ataxia

694. All may be the features of multiple sclerosis except:
   A. Spasticity
   B. Nystagmus
   C. Mental or emotional disturbances
   D. Vague somatic aches and pain

Ans.: 686. C   687. D   688. B   689. D   690. C   691. D   692. B
       693. D   694. D

695. All of the following diseases may be mistaken for multiple sclerosis except:

A. Porphyria

B. Guillain-Barre syndrome

C. Platybasia

D. Tuberculoma in temporal lobe

696. All of the following areas of brain are commonly involved in amyotrophic lateral sclerosis except:

A. Bulbar motor nuelei

B. Corticospinal pathway

C. Anterior horn cells

D. Posterior horn cells

697. All are the features of amyotrophic lateral sclerosis except:

A. Diffuse muscle atrophy with spastic paraparesis of lower limbs

B. Fasciculations

C. Retained reflexes in the muscles

D. Patehy sensory loss

698. "Useless hand" (Astereognosis, loss of sense of position, awkwardness in carrying out fine movements of fingers or difficulty in holding small objects) is seen in:

A. General paralysis of insane

B. Tuberculoma in temporal lobe

C. Multiple sclerosis

D. Dorsolumbar tuberculosis

699. The commonest cause of death in neuromyelitis optica (Devic's disease) is:

A. Bed sores

B. UTI

C. Respiratory infection

D. Respiratory failure

700. Which of the following is not true about neuromyelitis optica (Devic's disease):

A. Common in males

B. Commonest during adolescence

C. An interval of a few days to months between the attacks of blindness and paraplegia

D. CSF shows lymphocytosis and a rise in the protein content

701. Which of the following is the most frequent presenting symptom of schilder's disease (diffuse sclerosis) in adults:

A. Mental disturbances

B. Visual disturbances

C. Myoclonic jerks

D. Hemiplegia

702. The speed at which axon conducts an action potential is directly related to its :

A. Diameter

B. Length

C. Branching

D. Diameter of its dendrites

Ans.: 695. D   696. D   697. D   698. C   699. D   700. A   701. B
702. A

703. **Non-myelinated axons differ from myelinated in that they:**
   A. Are not capable of regeneration
   B. Are not associated with Schwann cells
   C. Lack nodes of Ranvier
   D. Are more excitable

704. **About brain death, all are ture except:**
   A. Isoelectric EEG
   B. Heart amd lungs may be functioning
   C. Patient may be revived
   D. Follows after heart or lung stops functioning

705. **Hydromyelia is fluid in space:**
   A. Extradural   B. Subdural
   C. Subarachnoid   D. Spinal canal

706. **Facudomuscular are seen in all except:**
   A. Pseudomuscular hypertrophy B. Motor neurone disease
   C. Polymyositis   D. Poliomyelitis

707. **Commonest calcified mass in suprasellar region in a child is:**
   A. Craniopharyngioma   B. Medulloblastoma
   C. Meningioma   D. Tuberculoma

708. **Not a complication of phenytoin:**
   A. Ataxia   B. Hirsutism
   C. Gum hypertrophy   D. Osteoporosis

709. **"Worm eaten" or frankly cystic putamen, opalski cells in thalamus, globus pallidus and substantia nigra, Fanconi sysdrome and a "Sunray" cataract is diagnostic of:**
   A. Hurler's disease   B. Han's schuller Christian disease
   C. Tay Sach's disease   D. Wilson's disease

710. **Which of the following is not true of craniopharyngioma:**
   A. Growth failure
   B. Normal born age
   C. Bitemporal hemianopsia
   D. Diabetes insipidus in over 75% of cases almost always seen

711. **Which of the following eye changes almost always indicate a CNS lesion :**
   A. Enlargement of blind spot on perimetry
   B. Optic strophy
   C. Field defects on perimentry
   D. Bilateral diplopia

Ans.: 703. C   704. C   705. D   706. A   707. A   708. D   709. D
        710. B   711. D

712. **Commonest cause of raised intracranial pressure in a child is:**
    A. Meningitis
    B. Tuberculoma
    C. Brain tumor
    D. Epitepsy

713. **Which of the following is not a cause of seizures on first day of neonatat life:**
    A. Birth asphyxia
    B. Narcotis withdrawal
    C. Phsnyiketonuria
    D. Tetsny

714. **Treatment of choice of salaam fits is:**
    A. Carbsnarepine
    B. Phenobarbitone
    C. Acetezolamide
    D. ACTH

715. **Which of the following antiepiteptic drugs is excreted as ketone bodies:**
    A Primidone
    B. Ethosuximide
    C. Mysoline
    D. Sodium

716. **Which of the following drugs may produce myasthenia syndrome:**
    A. Primidone
    B. Carbanmazepine
    C. Valproate
    D. Ethosuximide

717. **An increase in the head circumfrence in the first 3 months of life by more than whch of the following arouse the suspicion of hydrocephalusa:**
    A. 0.5 cm per fortnightly
    B. 1.0 cm per fortnightly
    C. 1.0 cm per monthly
    D. 1.0 cm per 3 monthly

718. **Which of the following drugs may be used to decrease the production of CSF:**
    A. Diaizepsm
    B. Sodium valproate
    C. Acetazolamide
    D. Haloperidol

719. **Rapid, inconstant fine tremor exaggerated by movement, cholea, athotosis and fluctuations in tone and hepatic coma are encountered in :**
    A. Multiple sclerosis
    B. Parkinsonism
    C. Wilson's disease
    D. Alcoholism

720. **Next to medulloblastoma, the commonest brain tumor in a child is:**
    A. Astrocytoma of cerebrum
    B. Astrocytoma of cerebellum
    C. Ependymoma of IV centricle
    D. Craniopharyagioma

721. **Doli's eye movements are seen in:**
    A. Normal conscious infants
    B. Intact brain stem
    C. Midbrain lesion
    D. Comatose patient

Ans.: 712. C   713. D   714. D   715. D   716. D   717. B   718. C
   719. C   720. B   721. B

722. **In a 5 year old child, which of the following is indicative of raised intracranial pressure:**
   A. Suture separation
   B. Macewen's or crackpot's sign
   C. Erosion of anterior clinoid process
   D. Increase in depth of pituitary fossa

723. **Which of the following is an early sign of raised intracranical pressure:**
   A. Bradycardia, hypotension   B. Bradycardia, hypertension
   C. Tachycardia, hypotension   D. Tachtcardia, hypotension

724. **Drug contraindicated in myasthenia gravis is:**
   A. Pyridostigmine   B. Neostigmine
   C. Quinine   D. Ephedrine

725. **Meralgia paraesthetica involves:**
   A. Axillary nerve   B. Sural nerve
   C. Median nerve   D. Lateral cutaneous nerve of high

726. **Commonest intramedullary spinal tumor is:**
   A. Chordoma   B. Meningioma
   C. Ependymoma   D. Oligodendroglioma

727. **A newer drug which is being tried for the treatment of cerebro-vascular insufficiency is:**
   A. Cyciendelate   B. Cinnarizine
   C. Pentoxifylline   D. Xanthinol nicotinate

728. **The classical diagnostic tried of mental deterioration, visual failure and paralysis is seen in :**
   A. Hans-Schuller-Christian syndrome
   B. Niemann-Pick disease
   C. Tay Sach's disease
   D. Gaucher's disease

729. **The following urinary changes may be seen in Wilson's disease except:**
   A. Massive generalized aminoaciduria
   B. Moderate phosphasturia and urie aciduria
   C. Glycosnria
   D. None of the above

730. **Drug of choice in West syndrome (Mental retardation with salaam fits) is:**
   A. Carbamazepine   B. Clonazepsm
   C. Sodium calproste   D. ACTH

Ans.: 722. B   723. B   724. C   725. D   726. C   727. C   728. C
     729. D   730. D

731. **Pseudotumor cerebri may be a side effect of:**
    A. Furazolidone
    B. Nalidixen acid
    C. Carbamazepine
    D. Frusemide

732. **In some patients, although walking is so difficult, they can be seen running for a bus quite well (Kinesia paradoxa) is typically seen in :**
    A. Wilson's disease
    B. Cerebellar tumor
    C. Parkinsonism
    D. Astasia – Abasia (Hysteria)

733. **Postencephalitie type of parkinsonism differs from idiopathic paralysis agitans that:**
    A. Tremors always precede rigidity
    B. Rigidity always precede tremors
    C. Both appear simultaneously
    D. None of the above

734. **The most important lesion in Parkinson's disease is in :**
    A. Locus coeruleus
    B. Substantia nigra
    C. Brain stem
    D. Internal capsule

735. **Most useful in the treatment of Syndenham chorea (Rheumstic chorea) is:**
    A. Penicillin
    B. Aspirin
    C. Bed rest followed by gradual mobilization
    D. Cjlor romazine

736. **Which of the following symptoms of Parkinsonism responds least to Levodopa:**
    A. Tremor
    B. Rigidity
    C. Akinesia
    D. Gait

737. **Sanger Brown's ataxia differs from Friedreich's ataxia by the following except:**
    A. Later age of onset
    B. Occurrence of ptosis, eptic atrophy, ophthalmoplegia
    C. Spasticity in lower limbs
    D. Nystagmus

738 **All of the following are common precipitants of epilepsy except:**
    A. Hyperventillation
    B. Flickering light
    C. Overhydration
    D. Acidosis

739. **A Seizure arising in one motor cortex starts most frequently in any of the following except:**
    A. Thumb
    B. Eyclid
    C. Angle of month
    D. Great toe

Ans.: 731. B   732. C   733. B   734. B   735. C   736. A   737. D
738. D   739. B

740. **Most common type of meningoencephalocele is:**
   A. Frontal
   B. Temporal
   C. Parietal
   D. Subacute sclerosing panencephalitis
741. **Distinctive ECG showing bursts of triphasic slow waves, is often seen in:**
   A. Beningn lymphocytic meningitis
   B. Hypoglycemia
   C. Postictal phase
   D. Subacute sclerosing panencephalitis
742. **The contraindications of amantadine include all except:**
   A. Peptic ulcer   B. Severe renal disease
   C. History of convulsion   D. Pregnancy
743. **The contraindications of Praziquantel (a drug most useful in neurocysticercosis) include all except:**
   A. Pregnancy   B. Lactation
   C. Ocular cysticercosis   D. None of the above
744. **Idoxuridine is best indicated in:**
   A. Encephaliitis lethargica
   B. Acute lymphocytie choriomeningitis
   C. Herpes simplex encephalitis
   D. Subacute sclerosing panencephalitis
745. **Radiosotopic brain scan is used chicfty to localize:**
   A. Brain abscess   B. Subdural haematomas
   C. Supratentorial tumours D. Infratentorial tumours
746. **Sweaty feat syndrome is associated with:**
   A. G-6 P deficiency   B. Phenylalanine deficiency
   C. Branched chain ketonuria D. Isovaleric academia
747. **Macular vision and adult acquity is attained by a child at the ages (years) of:**
   A. 3, 5   B. 4, 6
   C. 5, 7   D. 6, 8
748. **Oculomotor nerve involvement is seen in all of the following except:**
   A. Cerebral aneurysm   B. Diabetes mellitus
   C. Disseminated sclerosis D. Hepatolenticular degeneration
749. **Following are the contraindications of ergotamine except:**
   A. Peripheral vascular disease B. Ischaemic heart disease
   C. Peptic ulcer   D. Pregnancies

Ans.: 740. D   741. D   742. D   743. D   744. C   745. C   746. D
       747. C   748. D   749. C

750. Ocular bobbing is usually seen in the lesion at :
   A. Occipital cortex   B. Midbrain
   C. Pons   D. Pituitary fossa

751. Periodic complex on EEG is seen in :
   A. Psychoses   B. Creutzfeldt-Jakob's disease
   C. Brain tumor   D. Normal relaxation

752. Generalized delta activity on EEG is seen in:
   A. Use of sedatives and hypnotics
   B. Major traquilizers
   C. REM sleep, dreaming
   D. Psychoses

753. EEG will not make a specific diagnosis in :
   A. Neurosis   B. Herpes simplex encephalitis
   C. SSPE   D. Status epilepticus

754. Photic stimuilation is used as an evocative technique for the induction of:
   A. Tonic clonic seizures   B. Petit mal (absence)
   C. Partial complex   D. Hysteric seizures

755. Sleep deprivation may be used as an evocative technique technique for seizure:
   A. Partial complex   B. Absence (petit mal)
   C. Grandmal   D. Hysteric

756. Hyperventilation may be best used as an evocative technique for seizure:
   A. Partial complex   B. Petit mal (absence)
   C. Grand mal (Tonic clonic)   D. None of the above

757. The following drugs induce rapid (beta) EEG activity except:
   A. Diazepam   B. Barbiturates
   C. Meprobamate   D. Chlorpromazine

758. Which of the following EEG activity may be generalized (not localized anteriority or posteriorly):
   A. Alpha   B. Beta
   C. Delta   D. All of the above

759. Which of the following variety of seizures is usually seen in childhood only:
   A. Tonic clonic   B. Partial
   C. Absence (peritmal)   D. All of the above

Ans.: 750. C   751. B   752. B   753. A   754. A   755. A   756. B
      757. D   758. C   759. C

760. Match the following:

Seizure

I. Imperical temporal lobe spike   (i) Tonic-clonic (grand mal)

II. Generealized 3 –Hz spike and   (ii) Partial complex (visual)
polyspike-and-wave

III Generalized spike and   (iii) Absence (petitmal)
polyspike-and-wave

IV. Occipital spike-and-wave   (iv) Partial complex (psychomotor)

761. Which of the following EEG pattern is most likely in brain tumor:
- A. Alpha activity
- B. Bifrontal beta activity
- C. Delta activity, phase reversal over left occiput
- D. Periodic complexes

762. Electrocerebral silence is diagnostic of:
- A. Normal relaxation
- B. Cerebral death
- C. Psychoses
- D. Brain tumor

763. EEG is useful in making a specific diagnosis:
- A. Cerebral tumor
- B. Cerebellar tumor
- C. Hepatic encephalopathy
- D. Cerebral abscess

764. A unilateral upper motor neurone lesion in internal capsule is characterized by:
- A. Muscle fasciculations
- B. VII Nerve palsy
- C. Ipsilateral hypotonicity
- D. None of the above

765. Internuncial neurons are essential part of:
- A. Withdrawal reflex
- B. Stretch reflex
- C. All reflexes
- D. None of the above

766. A patient with loss of function of posterior columns of spinal cord will exhibit all except:
- A. Normal plantar response
- B. A clumsy gait made worse on closure of eyes
- C. Partial loss of pain sensation
- D. Dimished vibration sense

767. Cereballar disease is characterized by all except:
- A. Intention tremor
- B. Intact muscle power
- C. Reduced muscle tone
- D. Loss of muscle joint sensation

768. Athetoid movements are found in lesion of :
- A. Cerebellum
- B. Thalamus
- C. Temporal lobe
- D. Basal gangtia

Ans.: 760. C   761. C   762. B   763. B   764. D   765. A   766. C
767. D   768. D

769. **Stimulation of post ganglionic sympathetic neurons causes all except:**
   A. Mydriasis
   B. Salivation
   C. Sweating
   D. Adrenaline secretion from adrenal medulla

770. **Emotional reactions affect the autonomic nervous system by acting at:**
   A. Cerebrum
   B. Hippocampus
   C. Hypothalamus
   D. Hypophysis

771. **A patient's head and eyes deviate to the left and the left arm extends immediately before a generalized tonic-clonic seizure develops. The possible origin of seizure is:**
   A. Diencephalon
   B. Left cerebral hemisphere
   C. Right cerebral hemisphere
   D. Cerebellum

772. **Which seizure usually does not follow head trauma:**
   A. Absence (petitmal)
   B. Partial complex, psychomotor variety
   C. Partial motor with Jacksontan match
   D. Partial motor with secondary generalization

773. **The EEG has all of the following chacteristics except:**
   A. Usually has a background activity in the alpha (8-13 Hz) range
   B. May be abnormal in psychologically disturbed people because of use of psychotropic medications
   C. May have abnormalities in 25 percent of clinically asymptomatic people
   D. Will be specially diagnostic in Gilles de in Tourette's syndrome

774. **Partial complex seizures (e.g. psychomotor seizures) compared to absence (petit mal) seizures) are:**
   A. Likely to disappear in young adult life
   B. Accompanied by automatisms
   C. Shorter in duration and apt to begin in childhood
   D. Likely to have an aura and postictal confusion

775. **Drug useful in absence seizures is:**
   A. Phenytoin
   B. Carbamazepine
   C. Phenobarbitone
   D. Valprole acid

776. **A 31 year old lady, who developed partial complex seizures with psychomotor symptomatology three years ago, has developed lethargy and confusion. The possible etiology is:**

Ans.: 769. D 770. C 771. C 772. A 773. D 774. D 775. D
776. C

A. Expansion of a temporal lobe tumor
B. Psychomotor status epilepticus
C. Antiepileptie medication intoxication
D. Development of a subdural hematoma from head injury.

777. **All of the following physical signs may indicate antiepileptic medication toxicity except:**
   A. Ataxia   B. Lethargy or stupor
   C. Aphasia   D. Dysmetria on heal shin testing

778. **Narcolepsy is usually associated with:**
   A. Generalized seizures   B. Fugue states
   C. Schizophrenia   D. Cataplexy

779. **The frequency with which a child's eylids blink during an absence is:**
   A. 13-30/sec.   B. 7-13/sec.
   C. 3/sec.   D. Highly variable

780. **The interictal EEG may be of diagnostic importance in seizure disorder:**
   A. Partial complex, psychomotor variety
   B. Tonic-clonic (generalized)
   C. Absence (Petitmal)
   D. All of the above

781. **A patient has loss of central vision in both eyes followed by throbbing, unilaterial headache. He is suffering from:**
   A. Amaurosis fugax   B. Classical migraine
   C. Tension headache   D. Delirium tremens

782. **A patient who complains of "seeing" and smelling garbage most likely has :**
   A. Partial complex seizures of temporal lobe origin
   B. Partial complex seizures of occipital lobe origin
   C. Tension headache
   D. Classical migraine

783. **Head injury may be aetiological cause of partial seizures in:**
   A. Childhood   B. Adolescence
   C. Middle age   D. All of the above

784. **All of the following may be the common causes of partial seizures in middle age except:**
   A. Glioblastoma   B. Metastatic bone tumor
   C. Cerebrovascular accident   D. Perinatal cerebral hypoxia

Ans.: 777. C   778. D   779. C   780. D   781. B   782. A   783. D
   784. D

785. Which of the following is a cause of partial seizures beginning in childhood :
A. Hysteria
B. Sinusitis induced cerebral absence
C. Congenital cerebral malformation
D. Cerebrovascular accident

786. All are the common causes of partial (elementary or complex) seizures in adolescence (13-21 year) except:
A. Sinusitis induced cerebral abscess
B. Conversion reaction
C. Mesial temporal sclerosis
D. None of the above

787. All of the following are hard neurologic signs except:
A. Difficulty with tandem gait
B. Right handedness at age one year
C. Asymmetric thumb size
D. Mirror-writing

788. Aphasis may be produced by lesion:
A. Cervical cord injury   B. Kernicterus
C. Cerebral anoxia   D. Stroke

789. Bender-Gestalt test may be abnormal with:
A. Mental retardation or cerebral palsy
B. Anxiety
C. Minimal brain dysfunction
D. All of the above

790. The Wechsler Intelligence scale for children may indicate:
A. Infantile autism   B. Attention deficit disorder
C. Cerebral palsy   D. Normal Intelligence

791. Which of the following symptoms of minimal brain disfunction are responsive to chemotherapy:
A. Dyslexia
B. Mirror movements or right-left confusion
C. Hyperactivity
D. All of the above

792. Which of the following types of cerebral palsy is most often associated with mental retardation:
A. Spastic quadriplegia   —   B. Spastic hemiplegia
C. Choreoathetosis   D. Mixed spastic chereoathetosis

Ans.: 785. C   786. D   787. D   788. D   789. D   790. D   791. C
792. C

793. **All of the following are soft neurologic signs except:**
   A. Unusual clumpsiness   B. Seizures
   C. Associated movements   D. None of the above

794. **Meissner's corpuscles are believed to be sensitive to:**
   A. Dopamine and histamine   B. Increase in pH of interstitial fluid
   C. Temperature fluctuation   D. Mechanical deformation

795. **Vomiting center is located in:**
   A. Hypothalamus   B. Pons
   C. Cerebral cortex   D. None of the above

796. **Which of the following is probably an inhibitory neurotransmitter:**
   A. Dopamine   B. Serotonin
   C. GABA   D. Norepinephrine

797. **Match the following:**
   I.   Choreoathetosis   (i)   Cervical cord injury
   II. Spastic hemiparesis   (ii) Cerebral anoxia
   III. Cortical blindness   (iii) Stroke
   IV. Spastic quadriplegia   (iv) Kernicterus
   A. I (i)   II (ii)   III (iii)   IV (iv)
   B. I (ii)   II (iii)   III (iv)   IV (i)
   C. I (iv)   II (ii)   III (iii)   IV (i)
   D. I (iv)   II (iii)   III (ii)   IV (i)

798. **Deafness may be caused by:**
   A. Stroke   B. Cerebral anoxia
   C. Kernicterus   D. Cervical cord injury

799. **In Positron Emission Tomography (PET) the following positron emitting radionuclides may be used except:**
   A. O_2
   B. CO_2
   C. ^{18}FDG ($^{18}F_2$ deoxyoglucose)
   D. None of the above

800. **Administration of above radionuclides give information regarding:**
   A. Oxygen update   B. Blood flow
   C. Glucose utilization   D. All of the above

801. **In Stroke, PET imaging is useful in acute studies to determine :**
   A. Aetiology
   B. Prognosis
   C. Viable from non-viable tissue
   D. All of the above

Ans.: 793. B   794. D   795. D   796. C   797. D   798. C   799. D
      800. D   801. C

802. In patients with seizure disorders, interictal 18FDG studies may show localized area of :
A. Increased glucose metabolism in interictal period
B. Decreased glucose metabolism during a seizure
C. Decreased glucose metabolism in and around a seizure with increased glucose metabolism during the seizure.
D. Any of the above

803. In In patients wht Huntington's chorea who have normal CT scans; may show in the striatum;
A. Increased glucose intake
B. Increased lactate intake
C. Decreased glucose intake
D. None of the above

804. Which of the following tissue is not visualized by N M R (Nuclear Magnetic Resonance) :
A. Gray Matter   B. White Matter
C. Soft tissue around skull   D. Skull (bone)

805. Brainstem auditory evoked responses (BAERs) normally consist of:
A. 3 waves   B. 5 waves
C. 7 waves   D. 9 waves

806. BAERs may be useful in the diagnosis of all except :
A. Multiple sclerosis
B. Small vascular lesions of brain stem
C. Central pontine myelinolysis
D. Glioma of cerebrum

807. Which of the following statement is incorrect about BAER and SERs (Somatosensory Eoked Responses):
A. They can be used to follow CNS function in comatose patient.
B. Can be used in uncooperative patients
C. They are unaffected by general anaesthesia and highdose barbiturates
D. They are not affected by inattention or drowsiness

808. If a patient collapses during lumbar puncture the following measures are useful except :
A. If needle has not been withdrawn, inject normal saline equal in amount to CSF withdrawn
B. If needle has been withdrawn, raise foot end of the sbed
C. Give dexamethasone or mannitol
D. None of the above

Ans.: 802. C   803. C   804. D   805. C   806. D   807. D   808. D

809. After lumbar puncture recumbency for which duration will prevent headache :
- A. 6 hours
- B. 12 hours
- C. 24 hours
- D. 1 week

810. In heat exhaustion, in contrast to heat cramps in mainly depletion of :
- A. Water
- B. Salt
- C. Magnesium
- D. All of the above

811. Drug useful in the treatment of heat hyperpyrexia is :
- A. Paracetamol
- B. Morphine
- C. Chlorpromazine
- D. Haloperidol

812. Following are the predisposing factors for heat hyperpyrexia except:
- A. Alcoholism
- B. Oldage
- C. Working in dry climate
- D. Heavy manual work in the presence of high temperature

813. In heat hyperpyrexia, following may be present except :
- A. Dry, hot and flushed skin
- B. Shallow breathing
- C. Tachycardia
- D. All of the above

814. Headache is due to :
- A. Traction of blood vessel wall
- B. Displacement of blood vessels
- C. Dilatation or irritation of wall of blood vessels
- D. All of the above

815. Which is the least common cause of bulbar paralysis :
- A. Poliomyelitis
- B. Encephalitis
- C. Acute infective polyneuritis
- D. Myasthenia gravis

816. The mainstay of treatment when paralysis results from acute infective polyneuritis (Guillain-Barre syndrome) is :
- A. Antibiotics
- B. Idoxuridine
- C. Steroids
- D. Decongestive therapy

817. Which of the following is incorrect findling in cerebral malaria :
- A. Seizures
- B. Inequality of pupils
- C. Neck stiffness and hyperreflexia
- D. CSF proteins raised but cell count in normal

Ans.: 809. C   810. A   811. C   812. C   813. B   814. D   815. D
816. C   817. D

818. **The risk of using intravenous quinine in the treatment of cerebral malaria is :**
    A. Seizures   B. Hypotension
    C. Arrythmias   D. Respiratory failure

819. **Cremesteric reflexes may be lost in all except :**
    A. Severe diabetic autonomic neuropathy
    B. Compressed fracture lower lumbar spine
    C. Sacral spinal cord injury
    D. Frontal meningiomas

820. **The daily dose of aminocaproic acid in a patient with cerebral haemorrhage is :**
    A. $1 - 2$ g   B. $4 - 8$ g
    C. $10 - 15$ g   D. $24 - 48$ g

821. **Which of the following decongestive measures is best in the treatment of hypertensive encephalopathy:**
    A. Frusemide
    B. Mannitol
    C. Dextran (low molecular weight)
    D. Steroids

822. **The venous thrombosis is most frequent during pregnancy and puerperium in :**
    A. Superior sagittal sinus   B. Inferior sagittal sinus
    C. Cortical vein   D. Petrosal sinus

823. **Antimicrobials which cross the blood brain barrier include :**
    A. Chloramphenicol   B. Sulfadiazine
    C. Penicillin   D. All of the above

824. **Which of the following is never used intrathecally :**
    A. Penicillin   B. Gentamicin
    C. Streptomycin   D. Sulfadiazine

825. **The Indication of intrathecal steroids in a; patient with tuberculous meningitis is :**
    A. Seizures   B. Hemiplegia
    C. Spinal block   D. Cerebral oedema

826. **In a patient with paralysis, physiotheraphy should be started from :**
    A. Ist day
    B. 3rd day
    C. Ist week
    D. 2nd week

Ans.: 818. B   819. D   820. D   821. A   822. A   823. D   824. D
     825. C   826. A

827. The head of an unconscious patient should be :
A. Elevated   B. In flat position
C. Below body level   D. Any of the above

828. The risk of cerebral embolism in post myocardial infarction period is maximum in :
A. 1st week   B. 2nd week
C. 3rd week   D. First 3 days

829. After internal capsule, the cerebral haemorrhage is most common in the region of :
A. Pons   B. Putamen
C. Temporal lobe   D. Hippocampus

830. The drug which has been used to prevent further bleeding in a patient with cerebral haemorrhage is :
A. Papaverine
B. Duvadilon
C. Nicotinic acid
D. Epsilon Amino Caproic Acid (EACA)

831. The cause of subarchnoid haemorrhage in a young patient of 20 years age can be :
A. Ruptured aneurysm   B. Brain tumor
C. Epilepsy   D. . A.V. malformation

832. Diazoxide used in the treatment of hypertensive encephalopathy requires precautions in a patient with :
A. Diabetes mellitus   B. Bronchial Asthma
C. Rheumatoid arthritis   D. None of the above

833. In Toxic metabolic coma, marked rigidity does not occure except in:
A. Diabetic ketosis   B. Dehydration
C. Acute hepatic failure   D. Renal tubular acidosis

834. The risk of aspiration pneumonia may be minimized by :
A. Aspirating the stomach contents before each feed to ensure the correct position of the tube.
B. Elevating the head of the patient during and for a short period after feed.
C. By restricting the volume of each feed to 200 ml.
D. All of the above

835. The most frequent presentation of cerebral thrombosis is :
A. Seizures   B. Hemiplegia
C. Hypotension   D. Unconsciousness

Ans.: 827. B   828. C   829. B   830. D   831. C   832. A   833. C
834. D   835. D

836. **All of the following signs will favour the diagnosis of intracerebral haemorrhage except :**
   A. Intense headache and rapid worsening of consciousness.
   B. Regular respiration
   C. Neck stiffnes
   D. Blurred disc margins.

837. **In intracerebral haemorrhage, lumbar puncture will be negative for first :**
   A. 8 – 10 hours
   B. 12 – 16 hours
   C. 16 – 24 hours
   D. 24 – 48 hours

838. **Which of the following drugs in never combined with oral anticoagulants in the treatment of cerebral thrombosis :**
   A. Aspirin
   B. Dipyridamole
   C. Sulfinpyrazone
   D. All of the above

839. **Match the following :**
   Lesion   *Pupil*
   I. Thalamus   Dilated and fixed
   II. Midbrain   Small but reacts to light
   III. Pons   Constricted and little reaction to light
   A. I (i), II (ii), III (iii)
   B. I (ii), II (i), III (iii)
   C. I (iii), II (ii), III (i)
   D. I (iii), II (i), III (ii)

840. **All of the following are true about eye movements in a comatose patient except :**
   A. Normally, sudden passive turning of the head to one side produces conjugate deviation of the eyes to the opposite side (doll's eye movements)
   B. Absence of oculocephalic reflex in a comatose patient implies thalamic dysfunction.
   C. Absence of oculocephalic reflex does not differentiate between structural and toxic metabolic pathology .
   D. Oculovestibular reflex also carries a similar interpretation.

841. **In a comatose patient, extension and abduction of all the four limbs (decerebrate state) indicate to ion in :**
   A. Thalamus
   B. Midbrain
   C. Pons
   D. Medulla

842. **In toxic metabolic coma, pupils are nealy always reacting to light except in poisoning with :**
   A. Glutethimide
   B. Atropine
   C. Heroin overdose
   D. None of the above

Ans.: 836. B   837. A   838. A   839. B   840. B   841. B   842. D

843. The antidote for LSD intoxication is :
A. Naloxone
B. Diuretics
C. Dextrose
D. Non-specific supportive therapy

844. Propranolol is used in the treatment of intoxication with?
A. Lead
B. Heroin
C. Mercury
D. Thyroxine

845. All of the following are autosomal dominant diseases except :
A. Wilson's disease
B. Huntington's chorea
C. Acute intermittent prarphyria
D. Familial Alzhemier's disease

846. Which of the following diseases have sex-linked recessive inheritance:
A. Adrenoleukodystrophy
B. Tuberous sclerosis
C. Hairy ears
D. Huntington's chorea

847. Pseudobulbar palsy is characterized by al all of the following except:
A. Increased jaw jerk
B. Frontal release signs
C. Atrophy of tongue
D. Dementia

848. Which of the following findings is found both in bulbar and pseudobulbar palsy:
A. Dysarthria
B. Depressed gag reflex
C. Emotional lability
D. Reduced voluntary palate movement

849. All of the following sign are present only in bulbar palsy and not in pseudobulbar palsy except :
A. Dysphagia
B. Increased jaw jerk
C. Decreased jaw jerk
D. Dementia

850. Which is the cause of slowly progressing dementia in an adolescent, along with ataxia and myoclonus;
A. Wilson's disease
B. Creutzfeldt-Jacob's disease
C. SSPE
D. Hepatic encephalopathy

851. Which of the following tests confirms the diagnosis of Alzheimer's disease :
A. CT scan
B. Pneumoencephalography
C. Cerebral angiography
D. None of the above

Ans.: 843. D   844. D   845. A   846. A   847. C   848. D   849. C
850. C   851. D

852. **Match the disease states with electrolyle disturbances that may be associated with changes in mental status :**

			Na	K	BUN
I.	Water intoxication	i.	152	4.9	58
II.	Dehydration	ii.	125	2.8	5
III.	Uremia	iii.	138	4.8	60

   A. I (i), II (ii), III (iii)   B. I (ii), II (i), III (iii)
   C. I (iii), II (ii), III (i)   D. I (iii), II (i), III (ii)

853. **Match the appropriate medications or therapy with intoxications**

   I. Bromide   i. Hypertonic saline
   II. Water   ii. Physostigmine
   III. Imipramine   iii. Non-specific supportive + sedatives
   IV. L.dopa   iv. Saline, diuretics
   A. I (i), II (ii), III (iv), IV (iii)   B. I (ii), II (iii), III (iv), IV (i)
   C. I (iv), II (iii), III (ii), IV (i)   D. I (iv), II (i) III (ii), IV (iii)

854. **Vitamin B$_6$ is used as an antidot for intoxication with :**
   A. INH   B. Organic phosphates
   C. L-dopa   H. A + C

855. **Which of the following sign is present in polymyositis :**
   A. Atrophy of muscles   B. Fasciculations
   C. Absent deep tendon reflexes D. Paresis of shoulders and hips

856. **Normal pressure hydrocephalus is characterized by all of the following except :**
   A. Dementia   B. Seizures
   C. Incontinence   D. Ataxic-apraxic gait

857. **CNS lupus is characterized by all of the following except :**
   A. Strokes   B. Psychosis
   C. Seizures   D. Incontinence

858. **Seizures may be the presentation of all of the following except;**
   A. Porphyria
   B. Tuberous sclerosis
   C. Wilson's disease
   D. Subacute Sclerosing Panencephalitis

859. **The neuropsychiatric presentations of porphyria include all of the following except;**
   A. Recurrent episodes of delirium
   B. Seizures
   C. Peripheral neuropathy
   D. Hysteria

Ans.: 852. B   853. D   854. D   855. D   856. B   857. D   858. C
   859. D

860. **Bromism is characterized by all of the following except :**
    A. Acneiform rash
    B. Psychosis
    C. Headache
    D. Peripheral neuropathy

861. **Dementia, Flapping tremors and rigidity are present in :**
    A. Obstructive hydrocephalus
    B. Arsenic poisoning
    C. Wilson's disease
    D. Myxedema madness

862. **A 60 year-old woman has aches and tenderness of the shoulder muscles, she is unable to lift her arms above her head. There is a blotchy red rash on her head and neck. The diagnosis is most likely:**
    A. Polyneuropathy
    B. Steroid myopathy
    C. Periodic paralysis
    D. Dermatomyositis

863. **The following tests may useful in the diagnosis of the above condition except :**
    A. Electromyogram
    B. Muscle enzymes
    C. Skin and muscle biopsy
    D. Nerve canduction velocities

864. **The following conditions may be associated with the above disease except :**
    A. Congestive heart failure
    B. Lung malignancies
    C. Gastrointestinal malignancies
    D. Polyarteritis nodosa

865. **A 32-year-old woman with chronic back pain has a sudden exacerbation while raking leaves. She has difficulty walking and pain that radiates from the low back down the right posterior thigh to the lateral ankle. She has paresis of plantar flexion of the right ankle and an absent ankle DTR. She has an area of hypalgesia along the right lateral foot. She has :**
    A. Herniation of $L_4 - L_5$ disc compressing T_{12} root on the right
    B. Herniation of $L_5 - S_1$ disc compressing T_{12} root on the left
    C. Herniation of $L_5 - L_1$ disc compressing S_1 root on the right
    D. Hysteria

866. **Naegleria infections of CNS are common in children or young adults who :**
    A. Eat pork
    B. Work in fields
    C. Swim in fresh water lakes
    D. Drink well water

867. **Naegleria infects CNS through :**
    A. Blood
    B. Olfactory nerves
    C. Vagus nerves
    D. Ascending infection from spinal cord.

Ans.: 860. D   861. C   862. D   863. D   864. A   865. C   866. C
867. B

868. **Which of the following finding in CSF is wrong in Naegleria infections :**
   A. Pleocytosis (neutrophils)
   B. RBC's and trophozoites in CSF
   C. Elevated protein content and low glucose
   D. None of the above

869. **Treatment of choice in Naegleria infections of CNS is :**
   A. Metronidazole
   B. Amphotericin – B
   C. Tetracycline
   D. Steroids

870. **The drug which has shown good response of Acanthamoeba infections of CNS is :**
   A. Metronidazole
   B. Amphotericin – B
   C. Sulfadiazine
   D. Tetracycline

871. **The most reliable method of diagnosis of tetanus is :**
   A. CSF analysis
   B. EEG
   C. Blood examination
   D. Clinical signs

872. **The most common cause of death in tetanus is :**
   A. Exhaustion
   B. Respiratory failure
   C. Convulsions
   D. Cardiac arrythmias

873. **Children and adults have the most common site of origin of cerebral emboli as :**
   A. Ear infection
   B. Heart disease
   C. Lung disease
   D. Blood dyscrasias

874. **The most common symptoms of cerebral embolism is :**
   A. Hydrocephalus
   B. Transient ischaemic attack
   C. Seizures
   D. Dementia

875. **Which of the following symptoms is least frequent in Vertebral artery insufficiency:**
   A. Vertigo and ataxia
   B. Drop attack
   C. Dysarthria and dysphagia
   D. Unilateral deafness

876. **MRI should be used with caution in patients with :**
   A. Acute stroke
   B. Brain stem lesions
   C. Demyelinating disease
   D. Cardiac pacemakers

877. **For regional cerebral blood flow (CBF) studies use inhalation or injection of :**
   A. O_2
   B. CO_2
   C. Xenon
   D. Ether

Ans.: 868. D   869. B   870. C   871. D   872. B   873. B   874. B
875. D   876. D   877. C

878. **Angiography remains unsurpassed in demonstrating :**
    A. Brain tumors
    B. Stenosis of small arteries and aneurysms
    C. Hemorrhage or deep infarction
    D. Cysts or tuberculomas

879. **In pneumoplethysmography and ophthalmodynomometry, instrument is placed against :**
    A. Cornea   B. Sclera
    C. Retina   D. Lens

880. **Which of the following finding of CSF is incorrect about meningococcal meningitis is :**
    A. Low CSF glucose
    B. High CSF Protein
    C. Polymorphonuclear leucocytosis
    D. Low Chloride

881. **Treatment of choice of meningococal meningitis is :**
    A. Penicillin   B. Sulfadiazine
    C. Tetracycline   D. Erthromycin

882. **The commonest cause of sporadic (nonepidemic) encephatitis is :**
    A. Japanese – B   B. Yellow fever
    C. Herpes simplex   D. KFD

883. **Herps simplex encephalitis has a predilection for :**
    A. Putamen   B. Frontal lobes
    C. Temporal lobes   D. Thalamus

884. **The chief long term effects of herps simplex encephalitis is :**
    A. Psychomotor seizures
    B. Intellectual deterioration
    C. Visual agnosias
    D. Disturbance in temperature regulations

885. **A young hypertensive housewife suddenly servere right retro-orbital pain. Prostration and a right third cranial nerve palsy. The most probable cause is rupture of aneurysm of :**
    A. Anterior communicating artery
    B. Posterior cerebral artery
    C. APosterior commonuicating artery
    D. Middle cerebral artery.

886. **All of the following seizures may be followed by a postictal psychosis:**
    A. Tonic-clonic (generalized)
    B. Partial elementary

Ans.: 878. B   879. B   880. D   881. A   882. C   883. C   884. A
      885. C   886. B

C. Partial complex

D. Partial motor with secondary generalization

887. A middle aged man has the sudden onset of the worst headache his life while watching television. Although he has nausea and vomiting, he is able to speake coherently to his wife. The most likely cause is :

A. Intracranial hemorrhage

B. Psychogenic headache

C. Migraine

D. Brain tumor

888. The medicines which may cause headache are all except :

A. Nitroglycerin

B. Reserpine

C. Oral contraceptives

D. Propranolol

889. Match the following :

Disease	*Type of headache*
I. Tic douloureux	i. Lethargy, papilloedema and generalized headache
II. Bell's palsy	ii. Mild headache, focal finding
III. Hypertensive encephalopathy	iii. Mastoid pain followed by nerve palsy
IV. Brain abscess	iv. Lancinating pain in the jaw

A. I (i), II (ii), III (iii), IV (iv)

B. I (iii), II (iv), III (i), IV (ii)

C. I (iv), II (ii), III (i), IV (ii)

D. I (iv), II (iii), III (ii), IV (i)

890. The precipitants of migraine include all except :

A. Stress

B. Menses

C. Redwine

D. Sleep

891. Migrainous headaches in children differ from those in adults that they are :

A. Bilateral

B. Non pulsating

C. Autonomic symptoms are rare

D. Visual auras and autonomic dysfunction may be the primary symptoms

892. The neurologic disorders which may give rise to visual hallucinations:

A. Migraine

B. Seizures from frontal lobe

C. Drug abuse (LSD etc.)

D. Alcohol withdrawal

893. In Lateral medullary syndrome, the following are involved except :

A. IX, X,XI nerves

B. XII nerves

C. Spinothalamic tract

D. Sympathetic tract

Ans.: 887. A   **888.** D   **889.** C   **890.** D   **891.** C   **892.** B   **893.** D

894. End plate potential :
    A. Is propagatory in nature   B. Follows all or none law
    C. Undergoes depolarization   D. All of the above

895. Jaw reflex is mediated by carnial nerve :
    A. V   B. VI
    C. VII   D. IX

896. The deficiency of which of the following trace element has been reported to result in muscular dystrophy (myopathy), growth failures, Keshan disease (manifesting as cardiomyopathy and Kaschinbeck disease (endemic human osteopathy).
    A. Copper   B. Zinc
    C. Inorganic lead   D. Selenium

897. Kluver – Bucy syndrome is characterized by all of the following except :
    A. Increased oral, sexual and aggressive behaviour
    B. Loss of visual recognition
    C. Decreased reactivity to visual stimuli
    D. Memory deficit.

898. Which of the following is more like a physiological investigation :
    A. X-ray skull
    B. CT Scan head
    C. Magnetic Resonance Imaging (MRI)
    D. Positron Emission Tomography (PET)

899. Romberg's sign (increased swaying) of the body due to dorsal column proprioceptive system) may occasionally be seen in psychiatric disorder :
    A. Schizophrenia   B. Mania
    C. Phobia   D. Hysteria

900. The ability to make use of intuition, sensitivity and awarness of subliminal cues to interpret clinical observations of individual and group patients is known as :
    A. Paranosic   B. Telepathy
    C. Third ear   D. Third nervous system

901. Progressive mental deterioration is unlikely in :
    A. Huntington's chorea   B. Sydenham's chorea
    C. Wilson's disease   D. None of the above

902. Which of the following is recessive sex-linked inheritance :
    A. Huntington's chorea   B. Sydenham's chorea
    C. Wilson's disease   D. None of the above

Ans.: 894. A   895. A   896. D   897. C   898. D   899. D   900. C
    901. B   902. D

903. **Which of the following is autosomal recessive :**
  A. Huntington's chorea   B. Wilson's disease
  C. Sydenham's chorea   D. None of the above

904. **All of the following movement disorders are bilateral except :**
  A. Athetosis   B. Chorea
  C. Hemiballismus   D. Tremors

905. **Which of the following disease is associated with mental retaradation and seizure :**
  A. Athetosis   B. Chorea
  C. Hemiballismus   D. All of the above

906. **Which of the following has intermittent movements :**
  A. Athetosis   B. Chorea
  C. Hemiballismus   D. Parkinsonism

907. **Parkinsonian tremors responds to :**
  A. L-dopa   B. Propranolol
  C. Amantadine   D. Any of the above

908. **All of the following phenomena occur in NREM sleep except :**
  A. Sleep walking   B. Bedwetting
  C. Night terrors   D. Angina

909. **Reading or language epilepsy most often resembles :**
  A. Absence attacks   B. Akinetic attacks
  C. Myoclonic epilepsy   D. None of the above

910. **The commonest cause of myoclonic seizures is :**
  A. Uraemia   B. Hepatic failure
  C. Creutzfeldt-Jakob disease D. Idiopathic

911. **The commonest sites of origin of complex partial seizures are :**
  A. Frontallobe + Amygdala   B. Hippocampus + Parietal lobe
  C. Hippocampus + Amygdala   D. Hypothalamus + Amygdala

912. **CSF in relation to blood is :**
  A. Hypertonic   B. Hypotonic
  C. Isotonic   D. B or C

913. **A 58-year-old woman has apathy and impaired ability to concentrate. Otherwise her history is unremarkable and neurologic examination is unrevealing. The next step in her evaluation is :**
  A. EEG
  B. Lumbar puncture
  C. CT Scan
  D. Additional preliminary examination

Ans.: 903. B   904. C   905. A   906. A   907. C   908. D   909. C
   910. D   911. C   912. A   913. D

914. **Which of the following is not true about akathisia :**
    A. Occurs early in course of treatment
    B. Improves when the medication is reduced
    C. Cannot be suppressed voluntarily
    D. Improves as patient remains on medication

915. **In morbidly obese persons, hypoventilation with consequent hypercapnia, hypoxia and finally somnolence is known as :**
    A. Kline Levine syndrome   B. Kluver Bucy syndrome
    C. Simmond's disease   D. Pickwickian syndrome

916. **Temporal lobe lesion is characterized by all except :**
    A. Illusions and hallucinations
    B. Korsakeoff like syndrome
    C. Unilateral anosognosia
    D. Affective or schizophrenic like phychoses

917. **Which of the following symptoms is uncommon in parietal lobe lesion:**
    A. Disturbed body schema   B. Postural sense deficits
    C. Speech disturbances   D. Mind blind syndrome

918. **Occipital lobe lesion is characterized by all except :**
    A. Visual anosognosia   B. Visual agnosias
    C. Illusions and hallucinations D. Kluver bucy syndrome

919. **Lesion of which of the following lobes is least likely to be misdiagnosed as hysteria or malingering :**
    A. Frontal lobe   B. Parietal lobe
    C. Temporal lobe   D. Occipital lobe

920. **Frontal lobe lesions are characterized by all except :**
    A. Memory deficits reflect a tendency to "forget to remember"
    B. Pseudodepressed syndrome
    C. Pseudoschizophrenic or Pseudomanic syndrome
    D. Mind blind syndrome

921. **Recent memory is stored in :**
    A. Parietal cortex   B. Hippocompus
    C. Thalamus and subthalamus   D. Hypothalamus

922. **Disorientation for persons is most often seen in :**
    A. Delirium   B. Dementia
    C. Epilepsy   D. Ganser's syndrome

923. **Match the following :**
    I.   Cerebral cortical anoxia   (i) Huntington's chorea
    II. Infarction of subthalamic nucleus (ii) Parkinson's disease

---

Ans.: 914. C   915. D   916. C   917. D   918. D   919. A   920. D
921. B   922. B   923. D

III. Depigmentation of substantia nigra (iii) Hemiballismus

IV. Atrophy of caudate nucleus   (iv) Myoclonus

A. I (i), II (ii), III (iii), IV (iv)

B. I (ii), II (i), III (iv), IV (iii)

C. I (iii), II (ii), III (i), IV (iv)

D. I (iv), II (iii), III (ii), IV (i)

924. **Cavitary lesions of the globus pallidus and putamen are characteristic of :**
   A. Choreoathetotic cerebral palsy
   B. Wilson's disease
   C. Creutzfeldt-Jakob's disease
   D. SSPE

925. **The movements in the following diseases may be ameliorated by neuroleptics except :**
   A. Huntington's chorea   B. Sydenham's chorea
   C. Wilson's diseases   D. None of the above

926. **Which of the following may develop in older children :**
   A. Huntington's chorea   B. Wilson's diseases
   C. Sydenham's chorea   D. All of the above

927. **All of the following are common spinal puncture reactions except :**
   A. Headache   B. Dizziness
   C. Weakness and anxiety   D. Depression

928. **Headache following spinal puncture is best controlled with :**
   A. Aspirin   B. Morphine
   C. Diazepam   D. Phenobarbitone

929. **Pellicle on the top of CSF collected in a tube indicates :**
   A. Bloody tap   B. Syphilis
   C. Tuberculous meningitis   D. Brain tumours

930. **After adequate antibiotic therapy, which CSF change tends to become normal first:**
   A. Cell count   B. Colloidal gold curve
   C. Serology   D. Pressure

931. **In Lange colloidal gold curve which type of curve has the maximum peak :**
   A. Normal   B. Tabetic
   C. Luteic   D. Paretic

932. **In all of the following conditions CSF pressure may be low except :**
   A. Multiple sclerosis   B. Diabetic coma
   C. Spinal cord tumor   D. Uremia

Ans.: 924. B   925. D   926. D   927. D   928. A   929. C   930. A
931. D   932. D

933. **In all of the following conditions, CSF pressure may be normal except :**
   A. Poliomyelitis
   B. Aseptic meningitis
   C. Spinal cord tumor
   D. TB meningitis

934. **A mino acid in CSF constitute what percentage of blood :**
   A. 10
   B. 30
   C. 50
   D. 70

935. **Which is the correct composition of CSF (all values are in meq/L) :**

	Sodium	Potassium	Calcium	Chlorides
A.	140	1-2	8-10	500-600
B.	120	4-5	8-10	700-750
C:	140	3-4	2-3	120-130
E.	325	12-15	4-7	700-750

936. **"Bloody Tap" of CSF may be best used for :**
   A. Cultures
   B. Animal inoculations
   C. Cytological
   D. A+B

937. **The danger of cisternal puncture is an injury to :**
   A. Midbrain
   B. Pons
   C. Medula
   D. Cervicaal spinal cord

938. **For an urgent investigation, which of the following puncture for obtaining CSF is safest (especially in infants) :**
   A. Lumbar
   B. Cisternal
   C. Ventricular
   D. Combined

939. **Combined multiple punctures for obtaining CSF is most useful in identifying :**
   A. Infectious diseases
   B. Demyelinating diseases
   C. Spinal block
   D. Brain tumours

940. **The following diseases may cause pseudobulbar palsy except :**
   A. Alzheimer's disease
   B. Bilateral frontal lobe infarcts
   C. Encephalitis
   D. Botulism

941. **The reversible causes of dementia are all except :**
   A. Cerebral syphilis
   B. Hypertensive encephalopathy
   C. CNS lupus erythematosus
   D. Parkinson's disease

942. **The irreversible causes of dementia are all except :**
   A. Lacunar state
   B. Myoclonic epilepsy
   C. Mercury poisoning
   D. Pellagra

943. **Migrainous headache is characterized by all of the following except:**
   A. Unilateral pulsating
   B. Accoompained by autonomic disturbances
   C. Visual scotomata
   D. Anhidrosis

Ans.: 933. D   934. B   935. C   936. D   937. C   938. C   939. C
   940. D   941. D   942. D   943. D

944. The common precipitants of migraine in a female is :
A. Menses
B. Oral pills
C. Stress
D. All of the following

945. Migrainous headaches differ from those in children from those in adults except :
A. Visual auras
B. Autonomic dysfunction
C. Visual hallucinations
D. None of the above

946. Which of the following diseases may ready in both bulbar and or pseudobulbar palsy :
A. Poliomyelitis
B. ALS (Amyotrophic Lateral Sclerosis)
C. Alzheimer's disease
D. Multiple sclerosis

947. All of the following diseases cause bulbar palsy except :
A. Poliomyelitis
B. Vertebrobasilar artery occlusion
C. Encephalitis
D. Myasthesia gravis

948. The common medications which may result in dementia are all except:
A. Insulin
B. Steroids
C. L-dopa
D. Ergot derivatives

949. Which of the following antihypertensives does not cause dementia :
A. Methyldopa
B. Nifedipine
C. Reserpine
D. Propranolol

950. The most common form of dementia accompanied by a perpheral neuropathy is :
A. Acute intermittent porphyria
B. Aresenic poisoning
C. Wernicke-Korsakoff syndrome
D. Pellagra

951. All of the following are reversible causes of dementia except:
A. Uremia
B. Acute intermittent porphyria
C. Cryptococcal meningitis
D. SSPE (Subacute sclerosing panencephatitis)

952. Headache is a sequelae of all of the following except:
A. Brain tumour
B. Glaucoma
C. Subdural hematoma
D. None of the above

953. The commonest couse of mycotic aneurysm:
A. Cryptococcosis
B. Bacterial endocarditis
C. Meningococcal septicemia
D. Aspergillosis

Ans.: 944. D   945. D   946. B   947. C   948. D   949. B   950. C
951. D   952. D   953. B

954. **To distinguish AVM (arterio-venous malformation) from neoplasm, the best is:**
   A. Fundus
   B. EEG
   C. Cerebral angiography
   D. CT scanning

955. **Which of the following is incorrect:**

	Age group	*Common symptoms of AVM*
A.	Neonatal	Cyanosis and respiratory distress
B.	Infancy	Seizures and hydrocephalus
C.	Between older children and adults	Headache and subarachnoid hemorrhage
D.	None of the above	

956. **Lateral sinus thrombosis (common in infants and children) are usually secondary to:**
   A. Tumours
   B. Otitis media
   C. Head injury
   D. Hematologic disorders

957. **In Gradenigo syndrome, there is pain in the distribution of nerve:**
   A. II
   B. V
   C. VII
   D. X

958. **The "great anterior radicular artery" of Adamkiewwiez is usually found at :**
   A. $T_{12} - L_1$
   B. $L_1 - L_2$
   C. $L_3 - L_4$
   D. $L_4 - L_5$

959. **Most clinical spinal cord strokes affect which region :**
   A. Cervical
   B. Midthoracic
   C. Lumbar
   D. Sacral

960. **Which of the following investigations is contraindicated in pure spinal cord infarction:**
   A. Myelography
   B. Spinal angiography
   C. CT scanning
   D. MRI

961. **Foix -Alajounine syndrome (characterized by spinal cord necrosis) is the result of:**
   A. Anterior spinal artery occlusion
   B. Posterior spinal artery occlusion
   C. Thrombosis of spinal veins
   D. Spinal fractures

962. **The commonest cause of hematomyelia is:**
   A. Hypertension
   B. Anticoagulation therapy
   C. Injury
   D. Tumours

Ans.: 954. C   955. D   956. B   957. B   958. B   959. B   960. B
961. C   962. C

963. Dandy and Blackfan introduced the terms:
A. Communicating and non-communicating hydrocephalus
B. Active and arrested hydrocephalus
C. Obstructive hydrocephalus
D. Normal pressure hydrocephalus

964. In Alzheimer's disease, which of the following types of hydrocephalus is common:
A. Normal pressure hydrocephalus
B. Hydrocephalus exvacuo
C. Communicating hydrocephalus
D. Obstructive type

965. The commonest form of hydrocephalus is:
A. Obstructive B. Communicating
C. Normal pressure type D. Any of the above

966. Which of the following is the only known cause of oversecretion hydrocephalus:
A. Otitic type B. Choroid plexus papilloma
C. Basilar artery ectasia D. Transcortal arachnoiditis

967. Isoelectric EEG may be seen in:
A. Barbiturate intoxication B. Extreme hypothermia
C. Brain death D. All of the above

968. Which of the following brain lesion will characteristically produce EEG changes:
A. Meningiomas B. Pituitary adenoma
C. Slow growing astrocytoma D. Abscess temporal lobe

969. A normal EEG and brain scan together almost exclude all except:
A. Supratentorial tumors
B. Brain abscess (in temporal lobe)
C. Infratentorial tumors
D. All of the above

970. The following factors of vascular lesions such as cerebral infarct and intracranial hemorrhages are important in producing EEG changes except:
A. Location of lesion B. Size of lesion
C. Type of lesion D. All of the above

971. Triphasic waves on EEG may be seen in :
A. Hepatic encephalopathy
B. Uremic encephalopathy
C. Encephalopathy due to pulmonary disease
D. All of the above

Ans.: 963. A 964. B 965. A 966. B 967. B 968. D 969. C
970. C 971. D

972. In Alzeheimer's disease affecting cerebral cortex, EEG often shows the abnormality in the range of waves :
   A. Alpha
   B. Beta
   C. Theta
   D. Delta

973. EEG abnormalities are little or absent in :
   A. Myxoedema
   B. Encephalitis
   C. Wernicke – Korsakoff disease
   D. Creutzfeldt – Jakob disease

974. Evoked potential is the record of electrical activity produced by groups of neurons within the:
   A. Cerebrum
   B. Brain stem
   C. Thalamus
   D. All of the above

975. Following sensory systems are often used in recording of evoked response except:
   A. Visual
   B. Auditory
   C. Tactile
   D. Gustatory

976. The technique for recording evoked potential is known as averaging because :
   A. It is the average activity of EEG
   B. The process involves repeating the stimuli many times and recording the electrical activity during a certain brief interval following each stimulus.
   C. The response is the average of stimulation of more than one sensory system
   D. At that point EEG in negative.

977. Ankle tendon reflex duration may be prolonged by :
   A. Hypercalcemia
   B. Hypokalemia
   C. Hyponatremia
   D. Hypomagnesemia

978. Prolonged ankle tendon reflex duration may be seen in :
   A. Cushing's disease
   B. Pheochromocytoma
   C. Diabetes mellitus
   D. Acromegaly

979. Thickened nerves (Hypertrophic neuropathy) may be seen in all except :
   A. Leprosy
   B. Neurofibromatosis
   C. Amyloidosis
   D. None of the above

980. Tinnitus may be caused by :
   A. Salicylates
   B. Streptomycin
   C. Phenytoin
   D. All of the above

Ans.: 972. C   973. C   974. D   975. D   976. B   977. B   978. C
      979. D   980. D

981. **Which of the following changes in fundus indicates malignant hypertension :**
    A. Copper wire arteries B. Silver wire artery
    C. Cotton wool exudates D. Arterioles as fibrous cord

982. **Small, round white patches "Roth spots" often associated with small hemorrhage the disc may be seen in :**
    A. Neuroblastoma
    B. Subacute bacterial endocarditis
    C. Tuberous sclerosis
    D. Central retinal artery thrombosis

983. **Pseudotumor syndrome (especially as a complication of pregnancy) is seen in :**
    A. Caverous sinus thrombosis
    B. Aseptic or primary thrombosis of longitudinal sinus
    C. Straight sinus Thrombosis
    D. Empty sella syndrome

984. **Prolonged ankle tendon reflex duration may be seen in :**
    A. Sarcoidosis B. Hypothyroidism
    C. Obesity D. Mild anasarca

985. **Ankle tendon reflex duration may be prolonged by drugs :**
    A. Arsenic B. Gold
    C. Lead D. Bromides

986. **Ptosis may be feature of :**
    A. Myasthenia gravis B. Ocular myopathy
    C. III nerve paisy D. All of the above

987. **Thickened nerves may be seen in :**
    A. Diabetes mellitus B. Acromegaly
    C. Sarcoidosis D. All of the above

988. **In fundus, pale white mass on or very close to the nerve head appearing like a :"Small bunch of grapes" is seen in :**
    A. Neuroblastoma B. Tuberous sclerosis
    C. Septic retinitis D. Retinitis piamentosa

989. **Papilloedma is not as early feature in the tumours of :**
    A. Frontal lobe B. Patieial lobe
    C. Temporal lobe D. Occipital tube

990. **Optic atrophy may be produced by all except :**
    A. Lead B. Tobacco
    C. Quinine D. Mercury

Ans.: 981. C 982. B 983. B 984. D 985. D 986. D 987. D
988. B 989. A 990. D

991. The commonest cause of amblyopia is :
    A. Tobacco   B. Methyl Alcohol
    C. Ergot   D. Lead or arsenic
992. Match the following:
    Antiepileptics   *Half life (approx.)*
    I.   Sodium valproatte   i.   15 hours
    II. Ethosuximde   ii. 60 hours
    III. Clonazapan   iii. 24 – 48 hours
    IV. Primidone   iv. 8 hours.
    A. I (i), II (iii), III (iii), IV (iv)   B. I (ii), II (iii), III (iv), IV (i)
    C. I (iii), II (iv), III (i), IV (ii)   D. I (iv) II (iii) III (ii) IV (i)
993. In which of the following, EEG is usually normal :
    A. Delirium   B. Hepatic encephalopathy
    C. Peeudodementia   D. Complex partial seidures.
994. Pseudodementia usually requires initial treatment with :
    A. Piracetam   B. Imipramine
    C. Haloperidol   D. Phenytoin
995. Myoclonus occurs must often in :
    A. Alzheimer's disease B. Huntington's disease
    C. Parkinsonism   D. Creutafeldt – Jakob diseases
996. Which of the following is hallmark of communicating hydrocephalus:
    A. Memory impairment   B. Unsteadliness of gait
    C. Myoclonus   D. Nystagmus
997. In subcortical dementias, most evident features is :
    A. Ataxia   B. Tremors
    C. Memory impairment   D. Focal neutrological signs
998. In addition to memory impairment, most patients with multiinfarct dementia show:
    A. Aphasia   B. Agnosia and apraxia
    C. Disorientation   D. All of the above
999. In demonstration past strokes, which of the following in most sensitive:
    A. EEG   B. CT Scan
    C. MRI   D. Carotid angiography
1000.In Alzheimer diseases, EEG usually shows:
    A. Triphasic waves
    B. Polyspike pittern
    C. A slowing of backgroun rhythms
    D. 3 per second of back ground rhythms

Ans.: 991. A   992. A   993. C   994. B   995. D   996. B   997. C
      998. D   999. C   1000. C

1001.In most of the left handed persons the dominant hemisphere for speech and language is:
  A. Right   B. Left
  C. Both   D. 90% right +10% left.

1002.The disturbance in the blood flow of which of the following arteries most often produce aphasia:
  A. Left sylvian artery
  B. Right sylvian artery
  C. Intenal arcuate artery
  D. Anterior communicating artery

1003.Following are types of sensory aphasia except :
  A. Omissions   B. Literal paraphasia
  C. Dysprosody   D. Verbal paraphasia

1004.Following are types of motor aphasia except :
  A. Conduction aphasia   B. Dysprosody
  C. Total aphasia   D. None of the above

1005.The drug, which is believed to have a role in Vogt-Koyanagi – Harada syndrome is:
  A. Ampicillin   B. Streptomycin
  C. Idoxuridine   D. Corticosteroids

1006.Mollaret meningitis (Benign meningitis) is caused by :
  A. Herpes simplex type I   B. Epidermoid cyst
  C. Histoplasmosis   D. Not known

1007.In AIDS, the viral nucleic acid has been detected in :
  A. Neurones   B. Astroglia
  C. Macrophages   D. Oligodendroglia

1008.Most common cause of secondary fungal infection of CNS in AIDS is :
  A. Cryptococcus   B. Histoplasma
  C. Cryptosporidium   D. Coccidiomycosis

1009.The most common cerebral syndrome of viral etiology in AIDS is:
  A. Dementia   B. Depression
  C. Anxiety states   D. Epilepsy

1010.The definite diagnosis of cerebral lesions in AIDS is :
  A. EEG   B. CT scan
  C. MRI   D. Brain biopsy

Ans.: 1001. B   1002. A   1003. C   1004. D   1005. D   1006. D
     1007. C   1008. A   1009. A   1010. D

1011. The major adverse effect of drug 3'– azido –3– deoxythymidine (AZT) is :
    A. Seizures
    B. Bone marrow depression
    C. Hepatic toxicity
    D. Nephrosis

1012. "Progressive locomotor ataxia" is a term used for :
    A. Paretic neurosyphilis
    B. Tabes dorsalis
    C. Congenital neurosyphilis
    D. Meningosyphilis

1013. Match the following :

		Described	
I.	Bayle (1882)	i.	Congenital tabes
II.	Wilhelm Erb (1892)	ii.	Congenital dementia paralytica
III.	Clauston (1877)	iii.	Dementia paralytica or paretic neurosyphilis
IV.	Hemak (1885)	iv.	Spinal cord disorder tabes dorsalis

    A. I (iv), II (iii), III (ii), IV (i)
    B. I (ii), II (i), III (iii), IV (iv)
    C. I (iii), II (iv), III (ii), IV (i)
    D. I (iv), II (iii), III (ii), IV (i)

1014. Which of the following CSF changes in neurosyphilis is wrong :
    A. Pleocytosis (lymphocytosis)
    B. Gamma globulin increase
    C. Low glucose
    D. Increase in total CSF proteins

1015. Following are the criteria for treatment failure in syphilis except :
    A. Clinical signs of syphilis persist
    B. There is four fold rise in VDRL titre
    C. VDRL positive after one year in patient treated for primary syphilis
    D. None of the above

1016. In patients allergic to Penicillin, the other effective alternative is :
    A. Sulfadiazine
    B. Doxycycline
    C. Streptomycin
    D. Erythromycin

1017. Indications of lumbar puncture include all of the following except :
    A. To obtain pressure measurement and sample of CSF for examination
    B. To aid in therapy by the administration of spinal anesthetics and occasionally antibiotics
    C. To lower raised intracranial pressure
    D. To inject air as in pneumoencephalography

1018. The commonest cause of "dry tap" is :
    A. Improperly placed needle
    B. Spinal tumor
    C. Arachnoiditis
    D. Arnold chiari syndrome

Ans.: 1011. B   1012. D   1013. C   1014. C   1015. D   1016. B
       1017. C   1018. A

1019.The commonest cause of bloody tap is :
  A. Subarchnoid haemorrhage
  B. Transfixation of a meningeal vessel by needle
  C. Spinal injuries
  D. Coagulation disorders

1020.CT scan of head is useful in differentiating all of the following except:
  A. Epidural, subdural and intracerebral haemorrhages
  B. Tumors from aneurysm
  C. Cysts from tuber culomas
  D. Brain oedema from hydrocephal us

1021.Angiography, in comparison to CT scan, is most useful in localizing:
  A. Brain tumors   B. Abscesses
  C. Aneurysm   D. Hydrocephalus, obstructive type

1022.In carotid angiography, dye is best injected into :
  A. Common carotid artery   B. Internal carotid artery
  C. External carotid artery   D. Vertebral artery

1023.In carotid angiography under optimal conditions arteries upto what lumen diameter may be visualized :
  A. 0.001 mm   B. 0.1 mm
  C. 1.0 mm   D. 2.0 mm

1024.Ventriculography is best indicated in :
  A. Frontal lobe tumors   B. Hydrocephalus
  C. Cerebral infarction   D. Aneurysms

1025.Which radioactive isotop is regularly used for the visulization of cerebral tumors :
  A. Phosphorus   B. Chromium
  C. Technetium   D. Fluoride

1026.The consistency of demonstration of CNS lesions by radioactive isotopes is maximum dependent on the lesions :
  A. Size   B. Pathology
  C. Vascularity   D. None of the above

1027.Ultrasound can be used to show CNS :
  A. Cysticercosis   B. Brain tumors
  C. Anterolateral frontal atrophy D. Atrophy of occipital lobes

1028.Match the following :

Disease		*Type of headache*
I.	Subdural empyema	i. Temporal pain, blindness, high ESR
II.	Temporal arteritis	ii. Moderate headache, focal seizures' fever

---

Ans.: 1019. B   1020. C   1021. C   1022. A   1023. B   1024. B
1025. C   1026. C   1027. B   1028. B

III. Viral meningitis   iii. Prolonged dull type with confusion and insomnia

IV. Posttraumatic   iv. Generalized, nuchal rigidity

A. I (i), II (ii), III (iii), IV (iv)   B. I (ii), II (i), III (iv), IV (iii)

C. I (iii), II (ii), III (i) IV (iv)   D. I (iv), II (iii), III (ii), IV (i)

1029. Classical migraine usually occurs in the following age groups except:

A. Childhood   B. Adolesence

C. Miiddle age   D. Older age

1030. Common migraine usually occurs in the following age groups except:

A. Childhood   B. Adolesence

C. Miiddle age   D. Older age

1031. Cluster headache usually occurs in the age group:

A. Childhood   B. Adolesence

C. Miiddle age   D. Older age

1032. Temporal arteritis most commonly occurs in the age group :

A. Childhood   B. Adolesence

C. Miiddle age   D. Older age

1033. The headaches relieved by sleep include:

A. Classical or common migraine

B. Cluster headache

C. Temporal arteritis

D. All of the above

1034. The following headaches may awaken a patient from sleep except :

A. Migraine   B. Brain tumor

C. Subdural hematoma   D. Tension

1035. In what stage of sleep, migraine headaches begin :

A. REM   B. NREM-stage 1

C. NREM-stage 2   D. NREM-stage 3 or 4

1036. The laboratory finding in migraine include all of the following except:

A. High serum serotonin   B. Low urinary 5-HIAA

C. Paroxysmal EEG   D. None of the above

1037. The diseases of conditions that cost industry the largest number of people-hours is :

A. Sciatica   B. Lowbackache and headache

C. Cervical spondylosis   D. Abdominal pain

1038. Which of the following headaches follow family patterns :

A. Classical migraine   B. Common migraine

C. Cluster headaches   D. Psychogenic headaches

Ans.: 1029. D   1030. D   1031. C   1032. D   1033. A   1034. D

1035. A   1036. A   1037. B   1038. C

1039.Match the following :

I. Tension headache	i. Propranolol
II. Infrequently occurring classical migraine	ii. Methysergide
III. Frequent, severe common migraine	iii. Cafergot
IV. Cluster headaches	iv. Aspirin compounds

A. I (i), II (ii), III (iii), IV (iv)   B. I (i), II (ii), III (iv), IV (iii)
C. I (i), II (iii), III (iv), IV (ii)   D. I (iv), II (iii), III (i), IV (ii)

1040.The drug used in the treatment of migraine associated with retroperitoneal fibrosis is:
A. Aspirin   B. Caffeine
C. Methysergide   D. Propranolol

1041.Match the following :

Deficit	*Artery of infarction*
I. Left VI and VII nerve palsy with hemiparesis	i. Left middle cerebral
II. Left palate paresis left limb ataxia.	ii. Perforating branch of basilar
III. Right hemiparesis with aphasia	iii. Vertebral or posterior inferior
IV. Left homonymous hemianopsia	iv. Right posterior cerebral

A. I (i), II (ii), III (iv), IV (iii)   B. I (ii), II (iii), III (i), IV (iv)
C. I (i), II (iii), III (iv), IV (ii)   D. I (iv), II (iii), III (ii), IV (i)

1042.A 66-year-old man substains a cerebrovascular accident (CVA) after which he is alrert but mute and unable to move his arms or legs. He has paresis of the palate, bilateral hyperrflexia and Babinski signs. He responds correctly to verbal and written questions by blinking his eyelids. He has normal EEG. Which of the following possibility is this case is wrong:
A. He has locked-in-syndrome
B. Lesion is at the base of the pontomedullary brainstem
C. He has aphasia but no visual impairment
D. Cortical processes seem to be intact

1043.Which of the following is not true about thrombotic or embolic cerebrovascular accident :
A. Sudden painless right hemiparesis and mutism if left hemisphere involved

Ans.: 1039. D   1040. C   1041. B   1042. C   1043. C

B. EEG may be abnormal because of extensive cerebral damage

C. Patient will be able to read or write, if he were nto paralyzed

D. An element of pseudobulbar palsy may develop if there is bilateral cerebral.

1044. Dreaming is usually associated with :

A. Delta waves in EEG

B. Decreased threshold to extrinsic stimuli

C. Increase in duration with age

D. None of the above.

1045. Which of the following is not a soft neurologic sign:

A. Hyperreflexia of DTR's and its asymmetry

B. Mirror movements

C. Choreiform or synkinetic movements

D. Athetosis

1046. Which of the following is not a hard neurological sign :

A. Babinski sign   B. Pseudobulbar dysarthria

C. Dyslexia   D. Scissor gait

1047. A one-year-old girl has a stroke because of sickle cell disease. This results in mild right hemiparesis. The following may probably be the additional consequences except :

A. Aphasis   B. Spastic cereberal palsy

C. Stunted growth of right arm   D. Seizures.

1048. Adults with penetrating head injuries (e.g. Gunshot wounds) frequently have all the following except :

A. Paresis   B. Seizures

C. Mental retardation   D. Aphasia

1049. Which of the following is not usually found in children with cerebral palsy:

A. Seizures   B. Hyperactivity

C. Dyslexia   D. Dysarthria

1050. Which of the following is never found in children with mental retardation:

A. Hyperactivity   B. Seizures

C. Dyslexia or dysarthria   D. None of the above.

1051. The lesion causing ideational dyspraxia is usually in the :

A. Anterior half of the dominant hemisphere

B. Posterior half of the dominant hemisphere

C. Anterior half of the non-dominant hemisphere

D. Posterior half of the non-dominant hemisphere

Ans.: 1044. D   1045. D   1046. C   1047. A   1048. C   1049. B

1050. D   1051. B

1052.The most frequent type of apraxia (dyspraxia) is :
   A. Limb-kinetic type   B. Innervatory
   C. Ideomotor   D. None of the above

1053.Loss of automatic respiration with preserved voluntary breathing (ondine's curse) occurs with lesions of:
   A. Pons   B. Medulla
   C. Midbrain   D. Cerebrum

1054.Which of the following types of breathing is characteristic of pontine lesion :
   A. Cheyne-Stokes respiration   B. Cluster breathing
   C. Apneustic breathing   D. Ataxic or Bitot breathing

1055.Ping-pong gaze or periodic alternating (ie, repetitive smooth excursions of the eyes first to one side and then to the other, with 2- to 3-second pauses in each direction) may occur in all of the following except:
   A. Bilateral cerebral infarction B. Cerebellar hemorrhage
   C. Midbrain lesion   D. Intact brainstem

1056.Perinaud syndrome characterized by downward deviation of the eyes and may be accompanied by pupils that do not react to light, is seen in lesion of thalamus or :
   A. Cerebrum   B. Pons
   C. Midbrain   D. Hypothalamus

1057.Bilateral ocular bobbing (conjugate brisk downward movements from the primary position) may be seen in :
   A. Cerebellar hemorrhage
   B. Pontine tegmentum lesion or transtentorial herniation
   C. Metabolic encephalopathy
   D. All of the above

1058.Unilateral bobbing signifies lesion in:
   A. Midbrain   B. Pons
   C. Medulla   D. Cerebrum

1059.Hypoventilation may be seen in all except :
   A. Brain stem injury   B. Neuromuscular disorders
   C. Sedative drugs over dosage D. Hepatic failure

1060.The following are the currently accepted criteria for brain death:
   A. Apnea during 10-20 minutes of oxygenation
   B. Absence of cephalic reflexes with fixed pupils isolectric EEG.
   C. Conditions present for 30 minutes at least 6 hours after the onset of coma and apnea
   D. All of the above

Ans.: 1052. C   1053. B   1054. C   1055. C   1056. C   1057. D
      1058. B   1059. D   1060. D

1061.The commonest type of entrapment neuropathy at elbow is of nerve
- A. Median
- B. Radial
- C. Ulnar
- D. Anterior interosseus nerve

1062.Which of the following is the least common cause of bacterial meningitis :
- A. H.influenza
- B. Meningococcus
- C. Pneumococcal
- D. Staphylococcus

1063.The most common neurological presentation of Vogt Koyanagi-Harada syndrome is:
- A. Somnolence
- B. Ocular palsies
- C. Headache
- D. Neurosensory deafness

1064.Differential diagnosis of seizures include :
- A. Migraine
- B. Transient Ischaemic attack
- C. Hysteria
- D. All of the above

1065.All of the following statements about stroke are correct except :
- A. The most characteristic feature of a stroke is its temporal profile.
- B. The suddenness of syndrome development stamps the disorder as vascular
- C. Thrombotic episodes are usually gradual inonset
- D. Rapid reversal of the deficit may occur from ischaemic cause and never in hemorrahage

1066.The neurological dificit in a stroke reflects all of the following except:
- A. Location
- B. Size
- C. Type
- D. Cause

1067.Anomias (dysnomias) seen in occlusion of arterial branches to occipital lobe are most severe for :
- A. Objects
- B. Colours
- C. Relatives
- D. Places

1068.Involvement of which part of the following areas will produce Korsakoff's psychosis :
- A. Midbrain
- B. Hippocampus
- C. Hypothalamus
- D. Thalamus

1069.A patient, who developed inaccurate visually guided limb movements, has:
- A. Ocular ataxia
- B. Optic ataxia
- C. Truncal ataxia
- D. Nystagmus

1070.Balint's syndrome is seen in the lesions of :
- A. Frontal lobe
- B. Parietal lobe
- C. Occipital lobe
- D. Thalamus

Ans.: 1061. C   1062. D   1063. C   1064. D   1065. C   1066. D
1067. B   1068. B   1069. B   1070. C

1071. An elderly patient developed coma (but no fever) quadriplegia and cranial nerve palsies, CSF studies and CT scaning were normal. He had :
  A. Tubercular meningitis   B. Viral encephalitis
  C. Basilar syndrome   D. Diabetic ketoacidosis

1072. Among the following, the least common cause of stroke is :
  A. Atherosclerotic thrombosis   B. Embolism
  C. Hypertensive hemorrhage   D. Ruptured aneurysm

1073. Transient warning attacks in carotid middle cerebral disease consist of :
  A. Hemiplegia   B. Speech disturbances
  C. Heimparesthesia   D. All of the following

1074. Surgical decompression of intracerbral hemorrhage is indicated if there is risk of all of the following except:
  A. Vital structures of the medulla at risk
  B. Declining level of consciousness
  C. Ventricular enlargement or herniation in an operative candidate
  D. Hemorrhage in sites like putamen, thalamus

1075. All of the following are the causes of sensory ataxia except :
  A. Syringomyelia
  B. Disorders of parietal lobe
  C. Cerebello – pontine angle tumors
  D. Spinal tumors

1076. Carpal tunnel syndrome may be produced by all of the following except :
  A. Myxoedema   B. Osteoarthritis
  C. Palindromic rheumatism   D. Conn's syndrome

1077. The causes of cranial bruits in an adult include:
  A. Carotid artery stenosis
  B. Arterio – Venous malformations of cerebrum
  C. Angiomatous conditions in the orbit
  D. All of the following

1078. All of the following are causes of gradually worsening diplopia except:
  A. Pituitary tumour   B. Compression from aneurysm
  C. Thyrotoxic opthalmoplegia D. Phenytoin overdose

1079. The central (intracranial) causes of laryngeal paralysis include :
  A. Bulbar lesions   B. Tumours
  C. Vascular lesions   D. All of the above

Ans.: 1071. C   1072. D   1073. D   1074. D   1075. C   1076. D
      1077. D   1078. D   1079. D

1080.Horner's syndrome may be produced by all except :
   A. Brainstem tumours   B. Syringomyelia
   C. Carotid artery thrombosis D. Temporal lobe abscess

1081.Parasomnia may be seen in all except:
   A. Diabetic ketoacidosis
   B. Uremia
   C. Portal systemic encephalopathy
   D. Amphetamine overdose

1082.Benign intracranial hypertension (Pseudotumor cerebri) may be produced by:
   A. Heavy metals   B. Abrupt steroid withdrawal
   C. Vitamin A intoxication   D. All of the above

1083.The therapy with which of the following antibiotic may produce increased intracranial pressure in children:
   A. Pencillin   B. Chloramphenicol
   C. Erythromycin   D. Kanamycin

1084.The deficiency of all of the following vitamins may produce loss of memory except:
   A. B_1 B. B_2
   C. B_{12} D. Niacin

1085.Loss of the memory may be a feature of all of the following except :
   A. Hypothyroidism   B. Multiple sclerosis
   C. Tumors of corpus collasum D. None of the above.

1086.The following diseases may result in nasal regurgitation of fluid:
   A. Acute bulbar paralysis   B. Polymyositis
   C. Myotonia dystrophica   D. All of the above

1087.Acute bulbar paralysis may be produced by all except :
   A. Myasthenic crisis   B. Poliomyelitis
   C. Botulism   D. Dermatomyositis

1088.Paralysis of tenth cranial nerve may be produced by thrombosis of artery:
   A. Middle cerebral   B. Posterior cerebral
   C. Posterior communicating   D. Posterior inferior cerebellar

1089.The central causes of nystagmus include:
   A. Disseminated sclerosis   B. Syringomyelia
   C. Friederich's ataxia   D. All of the above

1090.Hypersomnia may be seen in all except :
   A. Hypopituitarism   B. Severe anemia
   C. Hypothyroidism   D. Hyperthyroidism

Ans.: 1080. D   1081. D   1082. D   1083. B   1084. B   1085. D
       1086. D   1087. D   1088: D   1089. D   1090. D

1091.Symptomatic narcolepsy may be seen in :
   A. Multiple sclerosis
   B. Encephalitis
   C. Increased intracranial pressure
   D. Angina pectoris

1092.Following endocrinal diseases may result in increased intracranial pressure:
   A. Primary hypoparathyroidism
   B. Addison's disease
   C. Pseudohypoparathyroidism
   D. All of the above

1093.Loss of memory may be caused by drugs except:
   A. Benzodiazepines
   B. Anticonvulsants
   C. Amphetamines
   D. Phenothiazines

1094.Acute bulbar paralysis produced by all except:
   A. Rabies
   B. Diphtheria
   C. Encephalitis
   D. Myotonia dystrophica

1095.All of the following may produce nystagmus except:
   A. Alcoholic polyneuritis
   B. Myasthenia gravis
   C. Dermatomyositis
   D. Botulism

1096.Acute paralysis of only one leg may be produced by:
   A. Thrombosis of paracentral artery
   B. Multiple scierosis
   C. Cauda equina lesion
   D. Hypochondriasis

1097.Chronic paralysis of one arm may be produced by all except:
   A. Meningioma
   B. Syringomyelia
   C. Motor neurone disease
   D. SLE

1098.Following are the causes of prolonged ankle tendon reflex duration except:
   A. Neurosyphilis
   B. Parkinson's disease
   C. Myasthenia gravis
   D. Hyperthyroidism

1099.Following drugs may cause prolongation of ankle tendon reflex duration by all except:
   A. Propranolol
   B. Quinidine
   C. Reserpine
   D. Quinine

1100.Ptosis may be a feature of :
   A. Horner's syndrome
   B. Tabes dorsalis
   C. Periodic paralysis
   D. Frontal tumours

Ans.: 1091. D   1092. D   1093. C   1094. D   1095. C   1096. D
   1097 D   1098. D   1099. D   1100. D

1101. Raynaud's syndrome may be produced by :
- A. Syringomyelia
- B. Ergot poisoning
- C. Dermatomyositis
- D. All of the above

1102. Thickened nerves may be seen in :
- A. Refsum syndrome
- B. Dejerine – Sottas disease
- C. Charcot – Marie – Tooth disease
- D. All of the above

1103. Tinnitus may be produced by :
- A. Temporal lobe epilepsy
- B. Vascular lesion of pons
- C. Migraine
- D. All of the above

1104. Vomiting (Gastric crisis) may be a prominent feature of :
- A. Multiple sclerosis
- B. Tabes dorsalis
- C. Cannabis intake
- D. All of the above

1105. All of the following are effective in central causes of vomiting except:
- A. Chlorpromazine
- B. Promethazine
- C. Metoclopramide
- D. Trifluopromazine

1106. Transient hemisphere attack (THA) a variant of TIA, occurs most often in the territory of artery :
- A. Anterior cerebral
- B. Middle cerebral
- C. Posterior cerebral
- D. Vertebro-basilar

1107. Transient hemisphere attack most commonly involves :
- A. Upper extremity
- B. Lower extremity
- C. Face
- D. Trunk

1108. The term hypertensive encephalopathy was introduced in 1928 by :
- A. Oppenheimer and Fishberg
- B. Clauston
- C. Glasgow
- D. Hemak

1109. Reye's syndrome in children has been hypothesized to be the result of frequent intake of :
- A. Aspirin
- B. Emetics
- C. Acetaminophen
- D. All of the above

1110. The organelle mainly involved in Reye's syndrome is :
- A. Nucleus
- B. Nucleolus
- C. Ribosomes
- D. Mitochondria

1111. Which of the following investigation may be abnormal in Reye syndrome :
- A. CSF
- B. EEG
- C. CT Scan
- D. None of the above

Ans.: 1101. D   1102. D   1103. D   1104. B   1105. C   1106. B
1107. A   1108. A   1109. D   1110. D   1111. B

1112. **"Watershed" infarction is a common complication of :**
   A. Hypotensives
   B. Diuretics
   C. I/V fluids
   D. Mild dehydration

1113. **Strokes in children differ from those in adults in three important ways except:**
   A. Predisposing factors
   B. Clinical evolution
   C. Anatomic site of pathology
   D. Composition

1114. **Match the following :**

		Infarct
I.	24 hours	i. Non-homogeneous decreased density lesion secondary to edema.
II.	1 week	ii. Liquefaction necrosis and infarct becomes homogeneous with defined margins.
III.	3 months	iii. Necrotic infarct is replaced by a cystic fluid cavity and the lesion with sharp margins has the homogeneous density of CSF.

   A. I (i), II (ii), III (iii)
   B. I (ii), II (iii), III (i)
   C. I (iii), II (i), III (iii)
   D. I (iii), II (ii), III (i)

1115. **EEG over an occluded vessel usually shows:**
   A. Epileptic discharges
   B. Slow-wave abnormality
   C. Triphasic waves
   D. Any of the above

1116. **Medical therapy of TIA (Transient Ischaemic Attack)**
   A. Aspirin
   B. Anticoagulation
   C. Pentoxyphyline
   D. All of the above

1117. **Which of the following is not useful in acute ischaemic stroke:**
   A. Mannitol
   B. Glycerol
   C. Aspirin
   D. Corticosteroids

1118. **Routine use of glucose infusions is acute ischaemic strokes carries the risk of :**
   A. Irreversible brain damage by glucose-induced intracellular lactic acidosis.
   B. Hyperglycemia
   C. Hypokalemia
   D. Seizures

Ans.: 1112. A   1113. D   1114. A   1115. B   1116. D   1117. D
   1118. A

1119. The most characterstic histologic change in viral diseases of CNS is:
  A. Inclusions bodies
  B. Diffuse loss of cortical neurons
  C. Mononuclear infiltrate generally located around blood vessels
  D. Glial nodules and neuronophagia

1120. "Status spongiosus" characterized by "bubbles and Holes" in the gray matter is seen in:
  A. Creutzfeldt diseases   B. Pick's disease
  C. Azheimer's disease   D. SSPE

1121. The cause of "Acute necrotizing encephalitis" is most probably:
  A. Neurosyhills   B. Herpes
  C. Measles   D. Japanese –B virus

1122. All of the following are true about SSPE except:
  A. Dawson in 1933 describes "inclusion body encephalitis" of which SSPE is a type
  B. It generally attacks children and is characterized by personality changes and characteristic bursts of high voltage activity on EEG
  C. Measles antigen has been found in neurons and glial cells of patients with SSPE by direct fluorescent antibody
  D. Measles virus has been cultured from the brain biopsy tissue of all cases with SSPE

1123. Inclusion body encephatitis is characterized by :
  A. Subacute encephalitis
  B. Loss of nerve cells
  C. Cowdry A inclusion bodies in neurons and glial cells
  D. All of the above.

1124. Intranuclear inclusion bodies are seen in:
  A. Rabies
  B. Hepers simples encephalitis
  C. Progressive Multifocal Leukoencephalopathy
  D. Tabes dorsalis.

1125. Kuru is characterized by :
  A. Cerebellar ataxia   B. Shivering tremor
  C. Dementia   D. All of the above.

1126. Which of the following is most frequently involved in fungal infection.
  A. Meninges   B. Spinal cord
  C. Brain stem   D. Cerebrum

Ans.: 1119. C   1120. A   1121. B   1122. D   1123. D   1124. C
      1125. D   1126. A

1127. **Cryptococcus neoformans (Torula histolytica) is primarily an infection of :**
    A. Lungs B. Gut
    C. CNS D. Skin

1128. **The term "Lacunae" is used for :**
    A. Exudates in fundus in papilloedema
    B. Inclusion bodies in degenerative CNS disease
    C. Small infarcts
    D. Paradoxical embolus lodging in cerebrum

1129. **Lacunae are most often seen in**
    A. Putamen B. Gray matter
    C. Thalamus D. Base of pons

1130. **After glioma, the most common primary brain tumor is :**
    A. Meningioma B. Medulloblastoma
    C. Acoustic nerve tumour D. Ependymoma

1131. **Inflammatory nature of temporal arteritis is indicated by all of the following except:**
    A. Fever B. Marked leukocytosis
    C. Increased ESR D. Anemia

1132. **The cranial nerves which are particularly liable to injury in basal skull fracture include:**
    A. I B. II
    C. III D. VI

1133. **The commonest cause of death after head injury is :**
    A. Epilepsy B. Countrecoup injuries
    C. Raised intracranial pressure D. Direct injury to viral centers

1134. **Cerebral perfusion pressure (CPP) is defined as ;**
    A. Mean blood pressure – Intracanial pressure
    B. $\dfrac{\text{Mean blood pressure – Intracanial pressure}}{2}$
    C. $\dfrac{(\text{Mean blood pressure – Intracanial pressure})^2}{2}$
    D. $\dfrac{(\text{Intracranian pressure – 2})^2}{2}$

1135. **Cerebral perfusion pressure is usually measured in units:**
    A. mm Hg B. Cm H_2O
    C. mm CSF D. Toss

1136. **Cerebral perfusion pressure below what is detrimental to nerve cells:**
    A. 100-120 B. 80-120
    C. 60-80 D. 40-60

Ans.: 1127. A 1128. C 1129. B 1130. A 1131. B 1132. D
    1133. C 1134. A 1135. D 1136. D

1137.Plateau waves are produeced due to :
   A. Increase in cerebral blood volumes
   B. Drecrease in cerebral blood volumes
   C. Increase in compliance
   D. Decrease in cerebral blood flow.

1138.The immunoglobulin in CSF and important in multiple sclerosis is:
   A. IgA   B. IgM
   C. IgG   D. B+C

1139.For several minutes to an hour after acture head injury, cerebral blood flow increases this is known as:
   A. Autoregulation   B. Luxury perfusion
   C. Compensatory perfusion   D. None of the above

1140.In Glasgow coma score for injury which of the following is given maximum score:
   A. Spontaneuous eye opening
   B. Obeying motor response
   C. Localizing motor responses
   D. Orientation to verbal reponse

1141.Delayed traumatic collapse (after a head injury) may be due to:
   A. Injury   B. Pain
   C. Emotional upset   D. All of the above

1142.Which of the following is wrong:
   X-ray absorptive value on CT scan
   A. Blood   70 units
   B. CSF   8 units
   C. Subdural fluid after hemorrahage 70 units
   D. CSF with Xanthochromia   6 units

1143.Which of the following match is wrong
   Meningitis   *Drug of choice*
   A. Pneumo or meningococcal   Penicillin
   B. H influenza   Ampicillin
                                      (+ Chloramphenical)
   C. Staph aurreus   Penicillin
   D. Enterobacteriacese   Gentamicin

1144.Type A (Cowdry) inclusion bodies are seen in:
   A. Herpes simplex encephalitis
   B. Cytomegalic inclusion virus
   C. SSPE
   D. All of the above

Ans.: 1137. A   1138. C   1139. B   1140. B   1141. D   1142. D
     1143. C   1144. D

1145. Which of the following is the only inclusion body which is diagnostic of viral disease of CNS:
    A. Type A Cowdry in SSPE
    B. Pick's body in Pick's disease
    C. Negri's body in Rabies
    D. Type B Cowdryin virul encephalitis

1146. Von Economo's disease is name given to :
    A. SSPE
    B. Postencephalitis Parkinsonism
    C. Herpes simplex encephalitis
    D. Progressive multifocal leukoencephalopathy (PML)

1147. Negri bodies are shaped:
    A. Oval
    B. Rounded
    C. Bullet shaped
    D. Any of the aove.

1148. Negri bodies are most frequently seen in :
    A. Parietal lobes
    B. Temporal lobes
    C. Thalamus
    D. Spinal cord

1149. In kuru, pathological changes are most marked in :
    A. Frontal lobes
    B. Hippocampus
    C. Cerebellum
    D. Medulla

1150. In which of the following diseases, inflammatory infiltrate is notably show:
    A. Progressive multifocal leukoencephalopathy
    B. Rabies
    C. St. Louis encephalitis
    D. Creutzfeld-Jakob diseases.

1151. Which of the following matching is wrong :

Hemorrhage	Most frequent cause
A. Epitural	Epilepsy
B. Subdural	Trauma
C. Intracerebral	Hypertension
D. Subarachnoid	Berry aneurysm

1152. A true hemorrhage is different from a hemorrhagic infact by all of the following except:
    A. Hemorrhage overlaps arterial supply
    B. Tissue architecture is destroyed
    C. It does not correspond to supply of a given artery
    D. None of the above

Ans.: 1145. C   1146. B   1147. D   1148. B   1149. C   1150. D
    1151. A   1152. D

1153. Epidural hemorrhage results commonly due to rupture of :
  A. Anterior cerebral artery
  B. Middle cerebral artery
  C. Middle meningeal artery
  D. Extradural venous sinuses

1154. The most frequent cause of subdural hygroma is :
  A. Trauma
  B. Viral infections
  C. Bacterial infections
  D. Hypertension

1155. The term "plaques Jaunes" refers to :
  A. Concussion injury of brain
  B. Old contusions
  C. Recent contusions
  D. Laceration

1156. The commonest cause of hematomyelia (blood inside the cord substance) is :
  A. Trauma
  B. Degenerative disease
  C. Infection
  D. Hypertension

1157. Punched out translucencies in skull may be produced by:
  A. Leukemias
  B. Sickle cell anemia
  C. Cushing's disease
  D. All of the above

1158. The most common cause of bloody CSF is :
  A. Hypertension
  B. Head trauma
  C. Bleeding from aneurysm
  D. Atherosclerosis

1159. The most common site of aneurysm in internal carotid artery is:
  A. Junction of middle and anterior cerebral
  B. Middle cerebral
  C. Posterior cerebral
  D. Its junction with posterior communicating artery

1160. Which of the following statement is wrong :

	Artery/circulation	Most frequent site of aneurysm
A.	Single aneurysms	Posterior circulation
B.	Anterior cerebral	Anterior communicating
C.	Middle cerebral	First major branching in Sylvian fissure
D.	Posterior aneurysms	Apical bifurcation of basilar artery

1161. First symptom of aneurysmal bleeding is:
  A. Disturbance in consciousness
  B. Headache
  C. Hemplegia
  D. Hemianopia

1162. The headache of aneurismal bleeding is characterized by all of the following except:
  A. Unilateral
  B. Occipital
  C. Localized
  D. Severe

Ans.: 1153. C   1154. A   1155. B   1156. A   1157. D   1158. B
    1159. D   1160. A   1161. B   1162. C

1163. Which of the following is pathognomic of subarachnoid hemorrhage:
   A. Kernig's sign
   B. Neck stkffness or low back ache
   C. Subhyaloid hemorrhage in elderly
   D. Subhyaloid or preretinal hemorrhage in adults

1164. Which of the following symptoms reflects aneurysm of the anterior cerebral artery:
   A. Hemiparesis   B. Aphasia
   C. Oculomotor palsy   D. Abulia or paraparesis

1165. Which of the following CSF findings is not common in subarachnoid hemorrhage:
   A. Elevated pressure   B. Xanthochromia
   C. Normal or high glucose D. Pleocytosis

1166. Which of the following rules out subarachnoid hemorrhage:
   A. Normal CT Scan
   B. No evidence of Xanthochrmia in CSF within first hour of bleeding
   C. Absence of RBC's and Xanthochromia in third week after hemorrhage
   D. Absence ofheurological signs (neckstiff, Kernig's sign, subhyaloid hemorrhage) and symptoms

1167. Investigation of choice in larga cerebral aneurysms:
   A. EEG   B. Cerebral angiography
   C. CT scan   D. Fundus

1168. Thr greatest risk of rebleeding in a patient with subarchnoid hemorrhage is in:
   A. First day   B. 2 – 7 days
   C. 8 – 14 days   D. 15 – 30 days

1169. Which of the following drugs has shown to reduce the incidence of persistent ischaemic deficts after subarachnoid hemorrhage:
   A. Prostacyclin   B. Naloxone
   C. Nimodipine   D. Kanamycin

1170. Which of the following lesions, the imaging capacity of CT scanning exceeds that of MRI (Magnetic Resonance Imaging):
   A. Lesions of white matter
   B. Differentiation of cystic lesions
   C. Demonstration of acute hemorrhage
   D. All of the above

Ans.: 1163. D   1164. D   1165. C   1166. D   1167. B   1168. A
        1169. C   1170. C

1171.CT scanning is superior to MRI in demonstration of all except:
   A. Secondaries at the base of skull
   B. Calcified lesions
   C. Fractures of skull and spine
   D. None of the above

1172.MRI is preferred over CT scanning in all of the following lesions except :
   A. Multiple sclerosis
   B. Acute hemorrhagic stroke
   C. Biswanger's disease
   D. Leukodystrophies

1173.MRI is particularly preferred over CT scanning in localizing the lesions of :
   A. Frontal lobe
   B. Temporal lobe
   C. Parietal lobe
   D. Brain stem

1174.In acute spinal cord syndromes, the investigation of choice is :
   A. X-ray spine
   B. Myelography
   C. CT scanning
   D. MRI

1175.Following are the limitations of MRI except:
   A. MRI incompatible tongs or halos for spinal traction
   B. Immobilized patient
   C. Patients who are unconscious on life support systems
   D. All of the above

1176.Myelography is more useful than MRI in:
   A. Intrinsic spinal cord lesions
   B. Suringomyelia
   C. Extramedullary lesions
   D. Disc prolapse or spondylosis

1177. In spinal cord injury investigation of choice is:
   A. Simple myelography
   B. CT scanning
   C. CT myelography
   D. MRI

1178.Investigation of choice in localizing epileptic foci is:
   A. EEG
   B. CT scanning
   C. MRI
   D. PET scanning

1179.In Single-Photon Emission Computed Tomography (SPECT), The radionuclides used emit:
   A. Alpha-rays
   B. Beta-rays
   C. Gamma-rays
   D. X-rays

Ans.: 1171. D   1172. B   1173. D   1174. D   1175. D   1176. D
      1177. C   1178. D   1179. C

1180. SPECT has been acknowledged as a potentially important diagnostic tool in differentiating:
   A. Primary and secondary tumors of brain
   B. Infarction and hemorrhage
   C. Dementia caused by Alzhiemer's and multiinfarct dementia
   D. Demyelinating diseases

1181. Nuclear Magnetic Resonance Spectroscopy (uses naturally occurring non-radioactive measurements of 13C, 23Na, 7Li, 31P an d1H) is important in studying brain:
   A. Anatomy
   B. Physiology
   C. Biochemistry and Metabolism
   D. None of the above

1182. Gummas in neurosyphilis have shown response to all except:
   A. Penicillin alone
   B. Penicillin +steroids
   C. Steroids only
   D. None of the above

1183. Retreatment in neurosyphilis is recommended if there is all of the following except:
   A. Progress of neurologic findings, especially if CSF pleocytosis persists
   B. If CSF cell count is not normal within six months
   C. If VDRL test in serum or CSF fails to decline or shows a four fold increase
   D. If the first course of treatment was supraoptimal

1184. The most common presentation of Lyme disease (erythema chronicum migrans, ECM) is :
   A. Fever
   B. Headache and stiff neck
   C. Convulsions
   D. Dementia

1185. The cranial nerve most frequently affected in Lyme disease is :
   A. V
   B. VI
   C. III
   D. VII

1186. Which of the following is abnormal finding in Lyme disease (ECM) with neurological findings
   A. CT scan
   B. EEG
   C. CSF pressure
   D. Oligoclonal bands

1187. Following drug has been found to be effective in the treatment of Lyme disease:
   A. Penicillin
   B. Tetracycline
   C. Ceftriaxone
   D. All of the above

Ans.: 1180. C   1181. C   1182. D   1183. D   1184. B   1185. D
1186. D   1187. D

1188. **Which of the following schistosomiasis has a special predilection for cerebral hemispheres:**
   A. S. japonicum
   B. S. haematobium
   C. S. mansoni
   D. All have equal

1189. **Which of the following is contraindicated in investigating CNS manifestation of Echinococcosis:**
   A. CT scan
   B. Angiography
   C. Ventriculography
   D. Brain biopsy

1190. **The following are the common presentations of neuro cysticercosis except:**
   A. Hydrocephalus
   B. Seizures
   C. Dementia
   D. Meningitis

1191. **Following are the CSF changes in meningitic form of neurocysticercosis except:**
   A. Raised pressure
   B. Elevated protein content
   C. Low glucose
   D. 50-100 cell count (polynuclear)

1192. **Praziquantel is most effective in the treatment of which form of neurocysticerosis:**
   A. Arachnoiditis
   B. Meningitic form
   C. Parenchymal form
   D. Calcified leading to hydrocephalus

1193. **Following is the most common neurological presentation of visceral larva migrans (toxocariasis)**
   A. Hydrocephalus
   B. Hemiparesis
   C. Damentia
   D. Depression

1194. **Drug useful in the treatment of toxocariasis is:**
   A. Thiabendazole
   B. Albendazole
   C. Diethylcarbazine
   D. All of the above

1195. **Regarding intracranial tumours which of the following statement is true:**
   A. Behavioural changes never precede neurological symptoms or signs
   B. Headache and vomiting occurs in all cases
   C. These conditions are not seen in children
   D. Primary tumors are more common than secondary tumours

1196. **Number of epileptic patients expected per thousand population:**
   A. 1-2
   B. 5-10
   C. 11-20
   D. 20-30

Ans.: 1188. A   1189. D   1190. C   1191. D   1192. C   1193. B
    1194. D   1195. A   1196. B

1197.The following are true about epileptic attacks excepts:
A. Sudden loss of consciousness
B. Self injury incontinence during attack
C. Tonic-clonic movements of the limbs
D. Attacks always occur due to stress factors

1198.The diagnosis of epilepsy in most of the instances is mady by:
A. Detailed clinic history collected from one or more eye witnesses
B. X-ray skull
C. EEG
D. CT Scan

1199.All of the following can cause epilepsy except:
A. Any focal lesion in the brain   B. Brain injury
C. Infections of the brain   D. Prolonged use of ganja

1200.Which of the following is not an indication for investigation in epilepsy:
A. Focal epilepsy
B. Presence of neurological deficits on examination
C. Febrite convulsion
D. Onset of fits above the age 30 years.

1201.Agrammatism is an important sign of a major lesion of the :
A. Sylvian fissure   B. Thalamic nuclei
C. Hipocampus   D. Normal Pressure

1202.At birth, the brain is about what % of its adult weight:
A. 25%   B. 40%
C. 60%   D. 75%

1203.The commonest cause of bacterial meningitis in children is :
A. Meningococcus   B. Pneumococcus
C. Streptococus   D. H.Influenzae

1204.Which of the following does not readily cross blood brain barrier:
A. Penicillin   B. Chloromycetin
C. Thiopentone   D. Sulfonamide

1205.Regarding brain development, which of the following has an excellent prognosis:
A. Hypoglycemia   B. Perinatal complications
C. Meningitis   D. Hypocalcemia

1206.The most common type of seizures in a neonate is :
A. Generalized tonic clonic   B. Absence attacks
C. Focal   D. Subtle

Ans.: 1197. D   1198. A   1199. D   1200. C   1201. A   1202. A
1203. D   1204. A   1205. D   1206. D

1207.Early sign of cerebral palsy is demonstration of:
- A. Moro's reflex at 2 months
- B. Moro's reflex at 5 months
- C. Rooting reflex at age of 2 months
- D. Palmer grasp at age 2 months

1208.Areas of CNS spared in poliomyelitis are the following except:
- A. Entire cerebral cortex leaving motor area
- B. Cerebellum except vermis and deep midline nuclei
- C. White matter of cord
- D. Gray matter of cord

1209.Neurological complications of typhoid fever include all of the following except:
- A. Encephalopathy
- B. Psychiatric complications like confusion, severe depression
- C. Guillain Barre syndrome
- D. Retinal detachment

1210.All of the following are the causes of sensory ataxia except:
- A. Peripheral neutitis
- B. Tabes dorsalis
- C. Disseminated sclerosis
- D. Freidreich's ataxia

1211.In syphilitic pachymeningitis, ataxia occurs due to effects on :
- A. Peripheral sensory nerves
- B. Posterior roots
- C. Posterior columns
- D. Labyrinthine apparatus

1212.Carpal tunnel syndrome may be seen in all except:
- A. Acromegaly
- B. Rheumatoid arthritis
- C. Addison's disease
- D. Primary amyloidosis

1213.All of the following are the causes of cranial bruits in an adult except:
- A. Glomus jugulare tumours
- B. Meningiomas
- C. Carotido-Cavernous sinus fistula
- D. Obstuctive hydrocephalus

1214.The following are the causes of temporary diplopia except:
- A. Myasthenia gravis
- B. Acute alcoholism
- C. Wernicke's encephalopathy
- D. Migraine

1215.Which of the following is not a feature of migraine:
- A. Headache
- B. Seizures
- C. Diplopia
- D. Dysphasia

1216.Parkinsonism like rigidity is seen with:
- A. Phenothiazines
- B. CO_2 poisoning
- C. Vertebrobasilar insufficiency
- D. All of the above

Ans.: 1207.B   1208. D   1209. D   1210. D   1211. B   1212. C
1213. D   1214. C   1215. B   1216. A

1217.After a week of head injury, a patient developed headache, apathy, clouding consciousness. He is most likely having:
   A. Chronic extradural heamatoma
   B. Chronic subdural haematoma
   C. Contusion injury
   D. Epilepsy

1218.Psamoma bodies may be found in:
   A. Oligodendroglioma   B. Meningioma
   C. Colloid cyst   D. Chromophobe adenoma

1219.Ulnar nerve palsy is characterized by all except:
   A. Loss of sensation on ulnar aspect of palm
   B. Loss of sensation in anatomical snuff box
   C. Weakness of interosset
   D. None of the above

1220.In a patient with cerebral compression, all of the following investigations are done except:
   A. EEG   B. Xray skull
   C. Lumbar puncture   D. CT scan

1221.Most useful investigation to rule out fracture of base of skull is:
   A. Xray skull   B. EEG
   C. Fundus examination   D. MRI

1222.All are possible in a patient who has extradural hemorrhage except:
   A. Blood found underneath the temporalis muscle
   B. Lucid interval is commonly found
   C. Most common site of bleeding is rupture of superior cerebral vein
   D. Disturbance in consciousness may be a present feature

1223.Best management of an infected subaponeurotic hematoma is :
   A. Repeated needle aspiration   B. Antibiotics alone
   C. Incision and drainage   D. Observation

1224.The most common site of intracerebral tumour in children is :
   A. Anterior fossa   B. Middle fossa
   C. Posterior fossa   D. Pituitary fossa

1225.Rhinorrhoea may be seen in fracture of:
   A. Cranial vault   B. Anterior fossa
   C. Middle fossa   D. Posterior fossa

1226.Anosmia may be seen in fracture of:
   A. Nasal bones   B. Cribriform plate of ethmoid
   C. Greater wing of sphenoid D. Base of skull

Ans.: 1217. B   1218. B   1219. B   1220. C   1221. A   1222. C
      1223. C   1224. C   1225. B   1226. B

1227.Peripheral neuropathy is seen in all except:
    A. Diabetes mellitus   B. Acute glomerulonephritis
    C. Chronic glomerulonephritis D. Amyloidosis

1228."Mad as a Hatter" refers to clinical picture caused by poisoning with:
    A. Arsenic   B. Mercury
    C. Cadmium   D. Selenium

1229.All of the following are physiological causes of punched out translucencies in skull except:
    A. Arachnoid granulations
    B. Emissary parietal foramina and parietal fenestrae
    C. Increased convulation-markings
    D. Hypocalcemia

1230.Following are the pathological causes of intracranial cacification except:
    A. Sturge-Weber syndrome   B. Tuberous sclerosis
    C. Meningioma   D. None of the above

1231.The occlusion of which of the following arteries is usually asymptomatic:
    A. Posterior cerebral
    B. Penetrating branch of middle cerebral
    C. Internal carotid
    D. Common carotid

1232."Abulia" or "Akinetic Mutism" may be produced by bilateral infarction of arteries:
    A. Anterior cerebral   B. Middle cerebral
    C. Posterior cerebral   D. Vertebro-basilar

1233.Pure motor hemiparesis is seen artery occlusion of which branch of middle cerebral:
    A. Main trunk   B. Upper division
    C. Lower division   D. Penetrating artery

1234.Weber syndrome or Benedikt syndrome may be seen in the lesion of:
    A. Hippocampus   B. Midbrain
    C. Pons   D. Medulla

1235.Ipsilateral facial palsy may be seen in the lesions of:
    A. Medulla   B. Pons
    C. Midbrain   D. None of the above

Ans.: 1227. B   1228. B   1229. D   1230. D   1231. D   1232. A
      1233. D   1234. B   1235. B

1236.Following are athe pathological causes of punched-out translucencies in skull except:
- A. Hyperparathyroidism
- B. Histiocytosis
- C. Myelomatosis
- D. Osteomalacia

1237.Common physiological cause of intractranial calcification in a person of 20 years age is calcification of:
- A. Choroid plexus
- B. Falx cerebri
- C. Pineal gland
- D. Hypophysis

1238.Which of the following is the unusual physiological intracranical calcification in an elderly is of:
- A. Choroid plexus
- B. Flax cerebri
- C. Hypophysis
- D. Pineal gland

1239.Which of the following is not an infective or infestive cause of intracranial calcification:
- A. Tuberculoma
- B. Cysticercosis
- C. Creutzfeld Jakob disease
- D. Hydatid cyst

1240.The term, "Reversible Ischemic Neurological Deficit" (RIND) is applied to syndromes that improve within:
- A. 12 hours completely
- B. 24 hours completely
- C. 24 hours but leave some minor neurologic abnormality
- D. Leave behind major neurologic deficit:

1241."Frontal gaze center" is an area:
- A. Left opercular cortex
- B. Prerolandic motor cortex
- C. Occipital cortex
- D. Pons

1242.Apractognosia is commonly seen in lesions of lobe:
- A. Right temporal
- B. Left temporal
- C. Right parietal
- D. Left parietal

1243.Following stain is used for demonstrating gliosis:
- A. Holtzer stain
- B. Cajal gole sublimate
- C. Marchi stain
- D. All of the above

1244.The term"torpedo" has been used for:
- A. Degeneration in astrocytes
- B. Axonal swelling in purkinje cells
- C. Cowdry type a inclusion
- D. Vacuolization of cell body.

1245.The "Red Neurone" is a name given to:
- A. Changes in brain tumor
- B. Acute neuronal injury
- C. Viral inclusion bodies
- D. Brain oedema

Ans.: 1236. D   1237. C   1238. C   1239. C   1240. C   1241. B
1242. C   1243. A   1244. B   1245. B

1246. Cowdry A inclusion bodies are seen in :
- A. Amyloidosis
- B. Alzheimer's disease
- C. SSPE
- D. Neurosyphilis

1247. Rosenthal fibres in brain are seen in:
- A. Neurosyphilis
- B. Syringomyelia
- C. Meningioma
- D. Secondaries in brain

1248. Which of the following type of herniation of brain is least common:
- A. Lateral or hippocampal herniation
- B. Outward herniation
- C. Downward herniation
- D. Upward types

1249. Following are the causes of hyperosmolar heperglycemic non-ketotic (HHNK) coma:
- A. Diabets mellitus or steroid therapy
- B. Burns
- C. Dialysis or immuno-suppressive therapy
- D. All of the above

1250. Drug which is used to localize dural hole is:
- A. Dianosil
- B. Radioactive gold
- C. Metrizamide
- D. Radioactive phosphorus

1251. Which of the following is the differential diagnosis of intracranial hypotension:
- A. Brain tumor
- B. Spinal block
- C. Metabolic toxicity
- D. Choroid plexus papilloma

1252. In hyperosmolar hyperglycemic non-ketotic diabetic coma, BUN-Creatinine ratio may be:
- A. 1 : 10
- B. 10 : 1
- C. 1:30
- D. 30:1

1253. All of the following constitue blood brain barrier except:
- A. Engothelial cells
- B. Oligodendroglia
- C. Basement membrane
- D. Astrocytic foot processes

1254. For demonstrating cerebral oedema dye used is :
- A. Metrizamide
- B. Conray
- C. Radioactive phosphorus
- D. Trypan blue

1255. Which of the following EEG activity may be generalized:
- A. Alpha
- B. Beta
- C. Theta
- D. All of the above

Ans.: 1246. C   1247. B   1248. B   1249. D   1250. C   1251. B
1252. D   1253. B   1254. D   1255. C

1256. Which of following EEG pattern is most likely with the use of sedatives or hypnotics:
    A. Alpha activity   B. Triphasic waves
    C. Bifrontal beta activity   D. Periodic complexes

1257. Normal relaxation is associated with the EEG pattern:
    A. Periodic complexes
    B. Triphasic waves
    C. Delta activity, phase reversal over left occiput
    D. Alpha waves

1258. Which of the following symptom or sign indicates a structural cerebral lesion:
    A. Absence attacks
    B. Psychomotor phenomenon
    C. Induction following hyperventilation
    D. Thrashing movements of extremities with pelvic thrusts

1259. Fortification scotomata is seen in:
    A. Classical migraine
    B. Common migraine
    C. Delirium tremens
    D. Amaurosis fugax

1260. A "shade of gray" covering one eye for three minutes is diagnostic of :
    A. Hysteria   B. Delirium tremens
    C. Amaurosis fugax   D. Classical migraine

1261. A patient sees a kaleidoscopic movement of bright lights in the right visual field may be seen in :
    A. Petit mal seizure
    B. Partial elementary seizure of occipital lobe
    C. Partial complex seizure of temporal lobe origin
    D. Delirium tremens

1262. Rapid extraocular movement artifact in EEG is seen in :
    A. REM sleep, dreaming
    B. Creutzfeldt-jakob's disease
    C. Hepatic encephalopathy
    D. Psychoses

1263. All of the following drugs may induce slowing of the EEG's background activity except :
    A. Lithium   B. Imipramine
    C. Chlorpromazine   D. Lorazepam

Ans.: 1256. C   1257. D   1258. B   1259. A   1260. C   1261. B
     1262. A   1263. D

1264. All of the following seizure-induced symptoms or signs suggest a structural cerebral lesion except :
A. Clonic movements of the left hand only
B. Jacksonian march
C. Incontinence
D. Uncinate seizure alone or before a partial complex seizure

1265. A patient with tremors sweating, tachycardia and complains of "seeing rodents" probably has :
A. Tension headache
B. Hysteria
C. Delirium tremens
D. Partial elementary seizures of occipital origin

1266. A patient complains of a red blotch of color in the left homonymous field followed by clonic movements of the left arm and leg, then the entire body. He most probably has :
A. Partial complex seizure of temporal lobe origin
B. Partial elementary seizure of occipital lobe origin
C. Partial elementary seizures of occipital origin with secondary generalizations
D. Classical migraine

1267. Urinary bladder has parasympathetic nerve supply:
A. $D_{12} - L_1$ B. $L_2 - L_4$
C. $L_5 - S_1$ D. $S_2 - S_4$

1268. Primary amyloidosis is characterized by neurologic findings:
A. Peripheral motor and sensory neuropathy
B. Peripheral neuropathy associated with cerebral manifestations
C. Guillain Barre type of syndrome
D. Spinal cord compression in thoracic region

1269. Posterior inferior cerebellar artery thrombosis is characterized by all of the following except:
A. Ipsilateral cerebellar ataxia and nystagmus to theside of lesion
B. Sudden onset of severe vertigo
C. Ipsilateral cervicosympathetic paralysis
D. Contralateral involvement of V and VIII nerves

1270. Commonest site of prolapsed intervertebral disc is :
A. $C_5 - C_6$ B. $C_2 - T_1$
C. $T_{12} - L_1$ D. $L_4 - L_5$ and $L_5 - S_1$

Ans.: 1264. C   1265. C   1266. C   1267. D   1268. A   1269. D
1270. D

1271.Papilloedema is characterized by all of the following except:
A. Marked loss of vision
B. Blurring of disc margin
C. Hyperemia of disc
D. Field defect

1272.All Horizontal movements of the eye are affected by lesion in the :
A. Midbrain
B. Cerebellum
C. Pons
D. Pretectum

1273.Injury to superior gluteal nerve gives rise to :
A. Loss of extension at hip joint
B. Loss of sensation over gluteal fold
C. Atrophy of gluteus maximus
D. Positive Trendelenburg teste

1274.The deep cerebellar nuclei:
A. Lie in the root of 4th ventride
B. Consist of dentate, fastigii, red nuclei
C. Related to lateral recess of 4th ventricle
D. Receive the afferents from cerebellum

1275.The reflex originating from golgi tendon organ to relax the responding muscle is a :
A. Monosynaptic reflex
B. Polysynaptic reflex
C. Disynaptic reflex
D. Reflex center of which is situated in medulla obiongata

1276.Hemiplegia is most often caused by thrombosis of artery:
A. Anterior cerebral
B. Middle cerebral
C. Posterior cerebral
D. Basilar

1277.A 8-year old girl was admitted with low grade fever, headache, vomiting and attacks of convulsions. On examination, no localizing signs were demonstrated. The CSF examination revealed Protein – 150 mg%, Sugar – 40 mg%, Chloride S20 mg% and cell count of 100/Cumm with large number of lymphocytes. The most likely diagnosis is :
A. Tuberculous meningitis
B. Pyogenic meningitis
C. Cerebral abscess
D. Viral encephalitis

1278.Ophthalmoplegia is a sign of all except:
A. Tangier disease
B. Leigh syndrome
C. Ataxia telangiectasia
D. Xeroderma pigmontosum

1279.An autosomal recessive disease, characterized by obesity, hypogonadism, mental retardation, micromelia, shortness of stature, hypotonia and lack of visual problems, is :

Ans.: 1271. A   1272. C   1273. D   1274. D   1275. C   1276. B
1277. A   1278. D   1279. C

A. Abstrom – Hallgren syndrome
B. Biemond syndrome
C. Prader – Willi syndrome
D. Laurence – Moon – Biedl syndrome

1280.The commonest supratentorial tumour in young adults is :
    A. Glioma
                B. Medulloblastoma
    C. Ependymoma
                D. Colloid eyst

1281.Not drains into cavernous sinus :
    A. Middle cerebral vein
           B. Ophthalmic vein
    C. Sphenoparietal sinus
           D. Great vein of Galen

1282.The weakness of small muscles of hand is seen in all of the following except:
    A. Motor neurone disease
         B. Syringomyelia
    C. Distal myopathy
            D. Diabetes mellitus

1283.Thickned nerves are seen in all except :
    A. Amyloidosis
             B. Refsum's disease
    C. Leprosy
              D. Sarcoidosis

1284.Erb's palsy is palsy of :
    A. Upper cervical cords
         B. Middle cerevical cords
    C. Lower cerevical cords
         D. C8 – T4 roots

1285.True about Wallerian degeneration is :
    A. Only axes cylinder remains for regeneration
    B. Only axon remains for regeneration
    C. Macrophage infilteration
    D. All of the above

1286.Myopathy is caused by drugs :
    A. Lithium
               B. Salbutamol
    C. Clofibrate
             D. All of the above

1287.Chronic alcoholism does not cause :
    A. Acute intoxication
        B. Peripheral neuritis
    C. Korsakoff"'s Psychosis
       D. Cardiomyopathy

1288.Peripheral neuropathy is not seen in :
    A. Diabets mellitus
        B. Amyloidosis
    C. Acute glomerulonephritis
    D. Chronic glomerulonephritis

1289.Not a cause of Raynaud's phenomenon :
    A. Lymphoma
    B. Scleroderma
    C. Systemic lupus erythematosus
    D. Cervical rib

Ans.: 1280. A   1281. D   1282. C   1283. D   1284. A   1285. C
      1286. D   1287. A   1288. C   1289. A

1290. Not a cause of raised intracranial pressure :
   A. Hypoventillation   B. Hyperventillation
   C. Hypothermia   D. Hypercarbia
1291. Spinal cord ends in intervertebral space of :
   A. $T_{12} - L_1$ B. $L_1 - L_2$
   C. $L_2 - L_3$ D. $L_3 - L_4$
1292. Injury to sensory cortex – 1 leads to loss of :
   A. Touch   B. Vibration
   C. Proprioception   D. All sensations
1293. Injury to medullary pyramid causes :
   A. Motor atrophy   B. Hyperalgesia
   C. Babinski sign   D. None of the above
1294. Carpal tunnel does not contain :
   A. Ulnar nerve   B. Profundus carpi
   C. Median nerve   D. Flexor carpi ulnaris
1295. Arousal response is through tract :
   A. Vestibulocortical   B. Tactospinal
   C. Retinculocortical   D. Rubrospinal
1296. The term, "Responsive mind trapped within an unresponsive body"
   is applied to :
   A. Alzheimer's disease   B. Neurosyphilis
   C. Guillain Barre syndrome   D. Parkinson's disease
1297. Postencephatitic Parkinsonism was first reported as a sequelae of
   virus :
   A. Influenza   B. Mumps
   C. Chickenpox   D. Rubella
1298. Which of the following is not involved in progressive supranuclear
   bulbar palsy:
   A. Globus pallidus   B. Subthalamic nuclei
   C. Substantia nigra   D. Frontal lobes
1299. Following cranial nerves nuclei are involved in Freidreich's ataxia
   except :
   A. VIII   B. X
   C. XI   D. XII
1300. Inherited peripheral neuropathy is seen in all except:
   A. Werdnig Hoffmann disease
   B. Charcot – Marie – Tooth disease
   C. Dejerine – Sottas disease
   D. Friedreich ataxia

Ans.: 1290. B   1291. B   1292. C   1293. C   1294. A   1295. C
      1296. D   1297. A   1298. D   1299. C   1300. D

1301. Alcoholic cerebellar degeneration is most marked :
- A. Flocculum
- B. Vermis
- C. Lateral cerebellum
- D. All of the above

1302. Dying back type of neuropathy is seen in :
- A. Subacute combined degeneration
- B. Diabetes
- C. Inherited redicular neuropathy
- D. All of the above

1303. Inhepatic encephalopathy, Alzheimer II astrocytes are found in all except:
- A. Lenticular nucleus
- B. Thalamus
- C. Substantia nigra
- D. Cerebellum

1304. Wilson's disease, cavitation is seen in all except:
- A. Lenticular nucleus
- B. Thalamus
- C. Red nucleus
- D. Mamillary bodies

1305. Absence of cerebral hemisphere, leaving only a thin, sac struture filled with fluid, is known as :
- A. Arrhinencephaly
- B. Porencephaly
- C. Hydranencehaly
- D. Anencephaly

1306. The term microcephaly is applied when an adult brain weighs less than :
- A. 1200 g
- B. 900 g
- C. 700 g
- D. 500 g

1307. The term megalocephaly is applied when brain weighs more than :
- A. 500 g
- B. 900 g
- C. 1500 g
- D. 1800 g

1308. The term "Etat Marbre" or "Status Marmoratus " is applied to hypermyelination seen in :
- A. White matter of cerebrum
- B. Basal ganglia
- C. Thalamus
- D. Spinal cord

1309. In Dandy Walker syndrome, there is agensis of :
- A. Frontal cortex
- B. Globus pallidus
- C. Vermis of cerebellum
- D. Corpus collasum

1310. Which of the following is wrong :

Period	Commonest cause of meningitis
A. Neonates	Esch. Coli
B. Infants and children	H.influenza
C. Adolescents and young adults	N.meningitides
D. Adults older than 40	Staphylococcus

Ans.: 1301. B   1302. D   1303. D   1304. D   1305. C   1306. B
      1307. D   1308. B   1309. C   1310. D

1311. The organisms most commonly implicated in subdural empyema are streptococci and :
    A. Staphylococcus aureus
    B. Esch. Coli
    C. H. influenza
    D. Gonococcus

1312. Which of the following is almost always observed in subdural empyema :
    A. Headache
    B. Focal neurological signs
    C. Neck stiffness
    D. Organism cultured from CSF

1313. Following are the CSF changes in asymptomatic syphilitic meaningitis except:
    A. Increase in white cells
    B. Normal or increase in protein content
    C. Positive serologic reactions
    D. None of the above.

1314. Major pathological finding in General Paresis of insane is :
    A. Diffuse loss of cortical neurons (Windswept cortex)
    B. Large numbers of hyperplastic glial cells
    C. Accumulation of lymphocytes and plasma cells around blood vessels.
    D. Iron positive granules in microglia.

1315. Following stain is used for demonstrating myelin sheath/myelin breakdown except :
    A. Nissl (Cresyl violet)
    B. Luxol – fast – blue
    C. Oil – Red – O Silver
    D. Woelcke stain

1316. The term " Dying back" is used for :
    A. Neuromal degeneration all at once
    B. Neuronal degeneration beginning at the synapse and moving along the axon towards the cell body.
    C. Neuronal degeneration beginning at cell body and moving towards synapse
    D. Neuronal degeneration beginning simultaneously in differents parts of a neurone

1317. The morphologic hallmark of nerve fibre injury or degeneration is :
    A. Lymphocytic infilteration
    B. Swollen or enlarge nerve fibre (spheroid)
    C. Fibroblasts
    D. Chromatolysis

1318. Alzheimer II astrocytes are seen in :
    A. Head injury
    B. Epilepsy
    C. Liver failure
    D. Creutzfeldt Jakob disease

Ans.: 1311. A   1312. C   1313. D   1314. A   1315. A   1316. B
       1317. B   1318. C

1319. The name "satellite cells" is given to :
    A. Astrocytes   B. Oligodendroglia
    C. Microglia   D. Ependyma

1320. Occlusion of foramina of Magendie and Luschka is seen in :
    A. Arnold Chiari syndrome
    B. Aneurysm of great vein of Galen
    C. Dandy – Walker syndrome
    D. Cyst in III ventricle

1321. Maximum incidence of calcification is seen is brain tumour :
    A. Medulloblastoma   B. Oligodendroglioma
    C. Meningioma   D. Craniopharyngioma

1322. J-shaped sella may be seen in all of the following except :
    A. Optic nerve ghoma   B. Empty sella syndrome
    C. Gargoylism   D. Low grade hydrocephalus

1323. All of the following can cause intracranial pathological calcification except :
    A. Carcinoma prostate   B. Cysticercosis
    C. Tuber culosis   D. Sturge Weber syndrome

1324. "Moth-eaten" appearance of skull is seen in :
    A. Syphillis   B. Tuberculosis
    C. Sarcoidosis   D. Neurofibromatosis

1325. Pituitary fossa may be enlarged in all of the following except:
    A. Nelson syndrome   B. Empty sella syndrome
    C. Acromegaly   D. Untreated cushing's syndrome

1326. Magnetic field strength used in magnetic resonance imaging is :
    A. Upto 2.5 Tesla   B. 2.5 – 5.0 Tesla
    C. 5 – 10 Tesla   D. More than 10 Tesla

1327. Spatial resolution of modern CT machine may be as low as :
    A. 0.5 mm   B. 0.1 mm
    C. 0.01 mm   D. 1 A

1328. Magnetic resonance imaging was first used by :
    A. Jack and Reese   B. Drayer and Roseabamn
    C. Damadian and Lauterbur D. Hounsfield

1329. Computed tomography (CT) was put to medical use first of all by :
    A. Jack and Reese   B. Damodian and Lauterbur
    C. Hounsfield and Carmack D. Drayer and Roseabamn

1330. CT scanning utilizes which of the following waves for image production:
    A. Gamma rays   B. Magnetic waves
    C. X-rays   D. Ultrasound waves

Ans.: 1319. B   1320. C   1321. D   1322. B   1323. A   1324. A
      1325. D   1326. A   1327. A   1328. C   1329. C   1330. C

1331.Which of the following does not show paraventricular calcification:
- A. Tuberculosis
- B. Herpes encephalitis
- C. Toxoplasmosis
- D. Tuberous sclerosis

1332.Classical CT appearance of extradural haematoma includes all of the following except:
- A. Biconvex in shape
- B. Seen as hyperdense area
- C. Concavo convex in shape
- D. Displaces lateral ventricle to contralateral side

1333.Earliest sign of raised intracranial pressure on X-ray skull is :
- A. Copper beaten or silver beaten appearance
- B. Sutural separation
- C. Erosion of floor of dorsum sellae
- D. Scalloping of inner table of skull

1334.Another 19 – year – old woman, who is adancer, presents with progressive weakness of the toes and ankles. On examination, she has only mild paresis in those areas, loss of only ankle reflex, unresponsive plantar reflex and decreased sensation in the toes and feet. She has :
- A. Myasthenia gravis
- B. Toxic polyneuropathy
- C. Polymyositis
- D. Thoracic spinal cord tumour

1335.Which tests would be most likely is to helpful in making a diagnosis:
- A. EEG
- B. Nerve conduction velocities
- C. Electronmyogram
- D. Tensillon test

1336.A 7-year-old boy is beginning to have difficulty standing upright. He has difficulty in standing up and also running. A cousin of the same age has a similar problem. The patient is well built and has a normal examination aside from paresis of his mucles and decreased quadriceps reflexes. He has :
- A. Peripheral neuropathy
- B. Duchene's muscular dystrophy
- C. Porphyria
- D. None of the above

1337.What test will help in diagnosing the above case :
- A. EEG
- B. Nerve conduction velocities
- C. Electromyogram
- D. Tensilon test

1338.What is the sex of his cousin :
- A. Male
- B. Female
- C. Intersex
- D. A or B

Ans.: 1331. A   1332. C   1333. C   1334. B   1335. B   1336. B
1337. C   1338. A

1339.Who is the carrier of the above conditions :
- A. Father
- B. Mother
- C. Either
- D. None of the above

1340.All of the following diseases may cause fasciculations except.
- A. Amyotrophic lateral sclerosis
- B. Insectiside poisining
- C. Fatigue
- D. Guillain – Barre syndrome

1341.All of the following are signs of peripheral nervous system dieases except
- A. Plantar reflexes absent
- B. Atrophic musles
- C. Spasticity
- D. Distal parents

1342.A 40-year-old watchmaker is unable to move his thumbs and finger. He had loss of touch of medical three fingers, but no change in reflexes. This most probable lesion is :
- A. Median nerve plasy
- B. Ulnar nerve plasy
- C. Radium toxically
- D. Lead intoxication

1343.All of the following paresis are common except :
- A. Ocular paresis
- B. Bulbar paresis
- C. Facical paresis
- D. Paresis of hands and feet

1344.In myasthenia gravis, which of the following tests is most likely to be helpful:
- A. Nerve conduction velocity
- B. Electromuyogram
- C. Tensilon test
- D. Muscle enzymes

1345.Which of the condition may be associated with myasthenia bravis except :
- A. Hyperthyroidism
- B. Thymoma
- C. Multiple scierosis
- D. None of the above

1346.To diagnose the asymptomatic carrier of Duchenne's musclar dystrophy, the test used is:
- A. EMG
- B. EEG
- C. Nerve conduction test
- D. Serum CPK level

1347.A 58-year-old man has difficulty walking lower back pains that radiate to the trunk and legs, loss of position sense (but intact pain and touch sense) in the feet, and pupils that extremities is normal but deep tendon reflexes are absent. He walks with a broad-based gait. What is the origin of the gait disturbance :
- A. Cerebellar disturbance
- B. Multiple sclerosis
- C. Spinal cord compressio
- D. Dysfunction of tracts of the spinal cord

Ans.: 1339. B   1340. D   1341. C   1342. B   1343. D   1344. C
1345. C   1346. D   1347. D

1348.Although thr pupils were small and unreactive to light, they were found to accomodate. This is known as :
A. Argyll Robertson pubil
B. Adie pupil
C. III Nerve palsy and VI nerve wealness.
D. Raised intractranial prossure

1349.While the patient has dysfunction of two parts of the nervouse system, he dis not have multiple sclarosis. The most likely diagnosis is :
A. Frontal lobe tumour
B. Tabes dorsalis
C. Syringomyellia
D. Fluorosis

1350.What laboratory test will be most usefull for the diagnosis of above.
A. CT scan
B. CSF VDRL
C. Nerve condution studies
D. Skeletal survey

1351.Which of the following is a reliable prognostic indicator of sequelae of head injury.
A. Retrograde amnesia
B. Anterograde amnesia
C. Psychogenic amnesia
D. Amnesia for remote events

1352.Beta waves in EEG are most prominent over :
A. Frontal lobe
B. Parietal lobe
C. Occipital lobe
D. Temporal lobe

1353.All of the following drugs may be useful in the treatment of tardive dyskinesia except:
A. Tetrabenazine
B. Pyrostigmine
C. Amphetamine
D. Lecithin

1354.In Wernicke's encephalopathy, the symptom which responds most promptly to treatment is :
A. Gait
B. Ocular movements
C. Dementia
D. Peripheral neuropathy

1355.Benzodiazepines if given in the first trimester of pregnancy, may produce:
A. Cleft lip
B. Atrial septal defect
C. Pulmonary stenosis
D. Respiratory distress syndrome

1356.Pseudopsychopathic and pseudodepressive syndromes are commonly associated with the disease of :
A. Frontal lobe
B. Temporal lobe
C. Parietal lobe
D. Occipital lobe

1357.The pleasure in gaining knowledge is known as :
A. Agromania
B. Essentialism
C. Epistemophilia
D. Ablutomania

Ans.: 1348. A   1349. B   1350. B   1351. B   1352. A   1353. C
1354. B   1355. A   1356. A   1357. C,

1358. **The drug of choice in hyperkinetic syndrome in children is :**
    A. Amphetamine   B. Imipramine
    C. Haloperidol   D. Diazepam

1359. **Which of the following conditions may lead to a syndrome characterized by delirium, ataxia and abnormalities of extraocular movements:**
    A. Hypoglycemia   B. Hyperthermia
    C. Myxoedema   D. Thiamine deficiency

1360. **Which of the following statements about CSF is wrong :**
    A. Specific gravity – 1.007   B. pH – 7.35
    C. Total base 157 meq/L   D. Chloride – 780 mg/100 ml

1361. **The brain tumours of which of the following lobes is most frequent cause of dementia:**
    A. Posterior frontal   B. Posterior temporal
    C. Anterior parietal   D. Anterior occipital

1362. **In dementing illnesses, the memory for recent events is lost before the memory for remote events (memory regression). This is called:**
    A. Retrograde amnesia   B. Anterograde amnesia
    C. Ribot's law   D. Mannkopf sign

1363. **In delirium, which of the following changes in EEG rhythm is most common:**
    A. From beta to alpha range   B. From alpha to beta range
    C. From delta to beta range   D. From alpha to delta range

1364. **In dementia, the other most frequent psychiatric illness is :**
    A. Delirium
    B. Schizophrenia
    C. Depression
    D. Obsessive compulsive neurosis

1365. **The drug of choice in a patient having dementia along with mild hypertension, elevated ESR, serum globulin abnormality, proteinuria and casts in urine :**
    A. Vitamin B_1 B. Piracetem
    C. Steroids   D. Thiazide

1366. **Neostigmine improves muscle weakness as it :**
    A. Blocks the action of acetylcholine
    B. Interferes with the action of amine oxidase
    C. Interferes with the action of choline acetyl transferase
    D. Interferes with action of acetylcholine esterase.

Ans.: 1358. A   1359. D   1360. D   1361. A   1362. C   1363. D
      1364. C   1365. C   1366. D

1367. Extensor muscle hyperreflexia in a patient with injury to midbrain between superior and inferior colliculi is due to :
   A. Decreased alpha efferent activity
   B. Loss of inhibition to gamma efferent neurons
   C. Generalized loss of facilitation
   D. Generalized loss of inhibition

1368. Following transient neurologic deficits are seen in the involvement of anterior (carotid) circulation except:
   A. Paresthesias of right arm and aphasia
   B. Amaurosis fugax
   C. Dysarthria
   D. Transient global amnesia

1369. Transient Global Amnesia (TGA) is seen in the distribution of :
   A. Anterior cerebral arteries
   B. Middle cerebral arteries
   C. Anterior communicating arteries
   D. Posterior cerebral artery

1370. Following neurologic deficits are seen in the involvement of distribution of posterior (basilar) circulation except :
   A. Fainting attacks
   B. Vertigo, nystagmus and ataxia
   C. Transient right hemiparesis without aphasia
   D. Aphasia

1371. A 70 year old man has five days of a left-sided headache, a non-fluent aphasia, right hemiparesis with hyperreflexia and a Babinski sign. He also has a right homonymous hemianopsia. The following could be the possible causes except:
   A. Carotid artery occlusion
   B. Brain abscess
   C. Subdural haematoma
   D. Brain tumour

1372. The excitatory postsynaptic potential (EPSP) differs from the end plate potential in that the EPSP is :
   A. Not associated with increased permeability to both sodium and potassium ions:
   B. Initiated by acetylcholine
   C. A reversal of charge
   D. Not decreased by d-tubocurare

Ans.: 1367. B   1368. D   1369. D   1370. D   1371. C   1372. D

1373.Which of the following symptom is rare in extra-axial lesions (masses outside of the brain stem e.g. subdural haematoma):
- A. Headache
- B. Hemiparesis
- C. Aphasia
- D. None of the above

1374.Match the following

Deficit	Artery of Infarction
I. Lower extremity monoparesis	i. Anterior spinal
II. Hemiparesis with relative sparing of leg	ii. Basilar
III. Quadriplegia with Intact position and vibration sensation	iii. Anterior cerebral
IV. Coma, quadriparesis	iv. Middle cerebral, either

- A. I (i), II (ii), III (iii), IV (iv)
- B. I (i), II (iii), III (ii), IV (iv)
- C. I (iii), II (iv), III (i), IV (ii)
- D. I (iv), II (iii), III (ii), IV (i)

1375.Right III nerve palsy with left hemiparesis is seen in the lesion of artery:
- A. Right posterior cerebral
- B. Perforating branch of basilar
- C. Left posterior cerebral
- D. Left middle cerebral

1376.Monoocular blindness from optic nerve ischaemia is seen due to involvement of artery:
- A. Vetebral
- B. Posterior inferior cerebral
- C. Ophthalmic
- D. Basilar

1377.Which of the following drugs may cause retrograde ejaculation:
- A. Atenolol
- B. Haloperidol
- C. Guanethidine
- D. Diazepam

1378.All of the following diseases may lead to internuclear ophthalmoplegia:
- A. Multiple sclerosis
- B. Lupus erythematosus
- C. Brainstem tumours
- D. All of the above

1379.The causes of cerebral palsy include all the of the following except :
- A. Congenital Rubella
- B. Werdnig-Hoffman disease
- C. Cerebral malformation
- D. Neonatal meningitis

Ans.: 1373. C   1374. C   1375. B   1376. C   1377. C   1378. D
1379. B

1380.The value of the EEG as a diagnostic test in children with suspected Minimal Brain Dysfunction (MBD) is :
A. It may help to exclude psychomotor seizures that can be confused clinically with MBD
B. There may be certain patterns that will indicate a diagnosis of MBD
C. A normal EEG tracing will exclude the diagnosis
D. None of the above

1381.Which disorders characterized by involuntary movements and dementia, may develop in adolescents :
A. Creutzfeldt-Jakob's disease B. Wilson's disease
C. Huntington's chorea   D. SSPE

1382.Dystonia musculorum deformans may be effectively treated with :
A. L-dopa   B. Neuroleptics
C. Tricyclic antidepressants   D. Stereotaetic ablation of portions

1383.Dystonia of the head and neck muscles may result from :
A. Wilson's disease   B. Cerebral palsy
C. L-dopa   D. All of the above

1384.Which of the following is not a hard neurologic sign:
A. Sustained clonus   B. Aphasia
C. Synkinetic movements   D. Chorea

1385.Which of the following is narcolepsy associated:
A. Cataplexy
B. Sleep paralysis
C. Sleep onset REM and hallucinations
D. All of the above

1386.Medications as well as particular illnesses cause impotence and other forms of sexual dysfunctions. In which illness is iatrogenic sexual dysfunction likely to be encountered except:
A. Psychosis   B. Migraine
C. Hypertension   D. Duodenal ulcer

1387.Retrograde ejaculation may result from:
A. Ovarian dysfunction
B. Diabetic autonomic neuropathy
C. Psychogenic cause
D. Sexual inexperience

1388.The neurologic illness which may cause sexual dysfunction are :
A. Parkinsonism   B. Poliomyelitis
C. Cerebral infarction   D. None of the above

Ans.: 1380. A   1381. A   1382. D   1383. D   1384. C   1385. D
1386. B   1387. B   1388. D

1389.The use of which of the following substances is not associated with optic neuritis:
  A. Tobacco   B. Oral contraceptives
  C. Ethyl alcohol   D. Methyl alcohol

1390.The preservation of temperature sensation despite loss of pain sensation indicates that:
  A. Lateral spinothalamic tract lesion
  B. Posterior columns are impaired
  C. Peripheral neuropathy
  D. None of the above

1391.Which of the following is not a cause of behaviour problem in a 6 year old girl who also has difficulty in climbing stairs and riding a tricycle.
  A. Minimal brain dysfunction   B. Cerebral diplegia
  C. Myasthenia gravis   D. Brain tumour

1392.Phosphotungstic acid hematoxylin (PTAH) stains neuroglial fibrils as:
  A. Red   B. Orange
  C. Blue or magenta   D. Pink

1393.Gliomatosis cerebri is :
  A. Neuronal degeneration   B. Inclusion cell encephalitis
  C. Cerebral abscess   D. A form of astrocytoma

1394.Juvenile Pilocytic astrocytomas are most frequent in :
  A. Cerebrum   B. Pons
  C. Cerebellum   D. Meninges

1395.Glioblostoma multiforme is most frequent in age group (years):
  A. 10 – 20   B. 20 – 30
  C. 30 – 40   D. Above 40

1396.Ependymoma (most frequent tumour of spinal cord) is most frequently seen in :
  A. IV ventricle   B. Cervical region
  C. Thoracic region   D. Lumbosacral region

1397."Myxopapillary ependymoma" is seen in:
  A. Thoracic cord   B. Lumbar cord
  C. Filum terminale   D. Cauda equina

1398.Antoni A and Antoni B are the terms applied to tumours :
  A. Glioma   B. Colloid cyst
  C. Ganglioglioma   D. Schwannomas

Ans.: 1389. B   1390. D   1391. C   1392. C   1393. D   1394. C
      1395. D   1396. D   1397. C   1398. D

1399."Shadow plaques" are seen in :
- A. Alzheimer's disease
- B. Pick's disease
- C. Multiple sclerosis
- D. Viral encephalitis

1400.Rosette formation is characteristically seen in tumour:
- A. Glioma
- B. Meningioma
- C. Cerebral neuroblastoma
- D. Ependymoma

1401.Neuroblastoma in brain arise most frequently in :
- A. Frontal lobes
- B. Temporal lobes
- C. Cerebellum
- D. Pons

1402.Medulloblastoma, in children, is most frequent in :
- A. Flocconodular region
- B. Lateral cerebellar hemisphere
- C. Vermis
- D. Restiform body

1403.Meningiomas arises from :
- A. Arachnoid cells
- B. Fibroblasts
- C. Blood vessels
- D. Any of the above

1404.Microglioma is a form of :
- A. Glioma
- B. Meningioma
- C. Lymphoma
- D. Secondary

1405.Schwannomas are most frequently seen in nerve:
- A. I
- B. II
- C. III
- D. VIII

1406.All of the following are true about MS plaques (seen in multiple sclerosis) except:
- A. Multiple
- B. Gray opalescent
- C. Irregular
- D. Seen only in white matter

1407.Which of the following conditions is associated with optic neuritis except:
- A. Multiple sclerosis
- B. Combined system disease
- C. Vasculitis
- D. Rubella

1408.Which of the following is an objective test for intelligence quotient in children :
- A. Rorschach
- B. Bender Gestalt
- C. Thematic Appreception Test
- D. Stanford Binet

1409.Which of the following may cause cerebral palsy:
- A. Blood group incompatibility
- B. SSPE
- C. Metachromatic Leukodystrophy (MLD)
- D. Brachial plexus birth traction

Ans.: 1399. C   1400. D   1401. A   1402. C   1403. D   1404. C
    1405. D   1406. D   1407. D   1408. D   1409. A

1410.The value of the EEG as a diagnostic test in children with suspected MBD (Minimal Brain Dysfunction) is :
   A. Triphasic waves indicate underlying cerebral tumour
   B. Abnormal EEG patterns may indicate some structural or physiclogical cerebral disorder.
   C. Normal EEG will exclude the diagnosis
   D. All of the above

1411.In the dominantly inherited form of dystonia musculorum deformans, which of the following is untrue:
   A. Patients tend to have involvement of the trunk and neck muscles first.
   B. Symptoms being in adult life
   C. Patients may present with tortipelvis
   D. CT scan shows cerebellar atrophy

1412.Spasmodic torticollis is :
   A. Confined to the muscles of neck and shoulder
   B. May be the first manifestation of dystonia musculorum deformans
   C. May be resisted by slight pressure applied to chin and often accompanied by retrocollis.
   D. All of the above

1413.Tourette's syndrome develops:
   A. Usually before 5 years of age
   B. Usually between 2 to 10 years of age
   C. After 12 years of age
   D. Girls are more affected

1414.In general, movement disorders are :
   A. Present intermittently 24 hours a day
   B. Absent during sleep
   C. Made worse by anxiety
   D. Suppressed for periods of upto 5 seconds by voluntary effort

1415.The following drugs may produce chorea except:
   A. Oral contraceptives   B. Reserpine
   C. L-dopa   D. None of the above

1416.Which of the following drug used in the treatment of hyperkinetic syndrome in children may produce chorea:
   A. Methylphenidate   B. Neuroleptics
   C. Antidepressants   D. All of the above

Ans.: 1410. B   1411. D   1412. D   1413. B   1414. A   1415. D
    1416. A

1417.Which of the following metabolic derangements may cause chorea except:
   A. Hypocalcemia   B. Hepatic failure
   C. Hyperthyroidism   D. Addison's disease

1418.A 25 year old woman had the recent onset of lethargy, temporal lobe seizures and lymphocytic pleocytosis of the CSF. Which of the following illnesses is most likely:
   A. Hypothyroidism   B. Schizophrenia
   C. Metastatic carcinoma   D. Herpes simplex encephalitis

1419.The following phenomenon occurs in NREM sleep except:
   A. Sleep walking   B. Cluster headache
   C. Bedwetting (Enuresis)   D. Night terrors

1420.The following phenomenon occurs in REM sleep except :
   A. Angina   B. Erections
   C. Nightmares   D. None of the above

1421.The indication of prednisolone in a patient with tubercular meningitis is :
   A. Seizures   B. Cranial nerve palsies
   C. Coma   D. Hemiplegia

1422.The basic pathology in a patient with post vaccinational encephalopathy is:
   A. Neuronal degeneration   B. Cerebral oedema
   C. Demyelination   D. Fibrosis

1423.The following drug should be used with care in a patient with acute viral encephalitis:
   A. Phenobarbitone   B. Glycerol
   C. Steroids   D. Low molecular weight dextran

1424.Drug of choice in post vaccinational encephalopathy is :
   A. Acyclovir   B. Vidarbine
   C. Steroids   D. Choloramphenicol

1425.The following drugs are useful in the treatment of herpes simplex encephalitis except:
   A. Cytosine arabinoside   B. Idoxuridine
   C. Vidarabine   D. None of the above

1426.The choice or combination of antiepileptic drugs depends on all except:
   A. Type of epilepsy   B. Severity of epilepsy
   C. Duration or cause of epilepsy D. Frequency of epilepsy

Ans.: 1417. D   1418. D   1419. B   1420. D   1421. B   1422. C
      1423. C   1424. C   1425. D   1426. C

1427. The cause of death in a patient with status epilepticus may be:
A. Dehydration
B. Systemic alkalosis
C. Cardiac arrythmias
D. Cerebral anoxia

1428. The causes or precipitating factors in status epilepticus include:
A. Associated infection
B. Electrolyte disturbances
C. Over-indulgence in alcohol
D. All of the above

1429. The most impotant aspect of management status epilepticus is :
A. Restart of antiepilepticus
B. Diazepam intravenously
C. Airway maintenance
D. Prevention from injuries

1430. The intravenous dose of phenytoin in the control of status epilepticus usually should not exceed:
A. 0.5 g
B. 1.0 g
C. 1.5 g
D. 2.0 g

1431. The intramuscular injections of anticonvulsants diazepam and phenytoin are not preferred because they:
A. Cause injection abscess
B. Cause muscle necrosis
C. Are absorbed erratically
D. Are not effective

1432. The intramuscular injection of which antiepileptic may often cause abscess:
A. Diazepam
B. Phenytoin
C. Phenobarbitone
D. Paraldehyde

1433. In an unconscious patient, most important in the care for :
A. Blood pressure
B. Airway
C. Diet
D. Hygiene

1434. Brudzinski's sign may be present in :
A. Meningitis
B. Poisioning
C. Diabetic ketoacidosis
D. Multiple sclerosis

1435. Patient who can be aroused only by intense stimulation such as loud shouts and deep pressure over the supraorbital notch on both sides has state of consciousness grade:
A. I
B. II
C. III
D. IV

1436. Cheyne-stokes respiration is usually associated with dysfunction of:
A. Thalamus
B. Midbrain
C. Pons
D. Cerebrum

1437. The commonest cause of Argyll Robertson pupil at present is:
A. Neurosyphilis
B. Multiple sclerosis
C. Diabetes mellitus
D. None of the above

Ans.: 1427. C   1428. D   1429. C   1430. B   1431. C   1432. D
1433. B   1434. A   1435. C   1436. A   1437. C

1438.Atypical Argyll Robertson pupil differs from typical or syphilitic type by that it:
 A. Responds to light
 B. Dose not respoind to accommodation
 C. Is pin point
 D. Is often dilated and unequal

1439.In all of the following conditions, CSF may be Xanthrochromic except:
 A. Brain tumor
 B. Spinal cord tumor
 C. Congenital toxoplasmosis
 D. Late CNS syphilis

1440.In alcoholism, which component of CSF may be altered (increased):
 A. Cell count
 B. Protein
 C. Glucose
 D. Colloidal gold

1441.Methyl alcohol has special affinity for cranial nerve:
 A. II
 B. III
 C. V
 D. VII

1442.Adenopathy may be a side effect of anticonvulsant:
 A. Diphenylhydantoin
 B. Ethosuximide
 C. Carbamazepine
 D. Valproate

1443.Unconsciousness in a patient is most commonly due to :
 A. Supratentorial mass lesions
 B. Subtentorial lesions
 C. Toxic causes
 D. Spinal cord injuries

1444.All of the following may be the causes of coma of gradual onset except:
 A. Brain tumor
 B. Meningitis
 C. Uremia
 D. Epilepsy

1445.Pandy test detects which component of cerebrospinal fluid:
 A. Albumin
 B. Globulin
 C. Chloride
 D. Calcium

1446.Pandy test is used to diagnose:
 A. Infectious diseases
 B. Degenerative disease
 C. Demyelinating diseases
 D. Epilepsy

1447.Peripheral neuropathy may be a side effect of all of the following except:
 A. Thiabendazole
 B. Colistin sulfate
 C. Furadantin
 D. None of the above

1448.Following drugs may produce CNS depression except :
 A. Ethosuximide
 B. Indomethacin
 C. Carbamazepine
 D. Prochloperazine

Ans.: 1438. D 1439. D 1440. A 1441. A 1442. A 1443. A
   1444. D 1445. B 1446. C 1447. D 1448. C

1449.Posterior superior iliac spine chosen as a bony mark for identifying intervertebral space for lumbar puncture is at the level of :
- A. $L_1 L_2$
- B. $L_2 L_{33}$
- C. $L_3 L_4$
- D. $L_4 L_5$

1450.Subdural tap is best done at which part of the Interal fontanelle :
- A. Medial most
- B. Lower most
- C. Upper most
- D. Lateral most

1451.CSF volume is maximem is :
- A. Lateral ventricles
- B. Subarachnoid spaces centrally
- C. Spinal subarchnoid
- D. III ventricle

1452.The word EEG is abbreviated for the :
- A. Electrolencephalograph
- B. Electroencephalography
- C. Electroencephalogram
- D. Any of the above

1453.EEG is the most important adjunct to :
- A. Clinical examination
- B. Blood analysis
- C. CT scan head
- D. PET scan

1454.Frequency in EEG is related to amplitude:
- A. Directly
- B. Are synonymous
- C. Inversally
- D. None of the above

1455.Which of the following is the most easily discernible rhythm on EEG:
- A. $0 - 4$ cycles / sec.
- B. $4 - 7$ cycles/ sec
- C. $7 - 13$ cycles / sec
- D. $13 - 20$ cockes/sec

1456.To activate latent abnormalities on EEG, the following techniques are commonly used except :
- A. Photic stimuli (rhythmic flashes of light)
- B. Hyperventilation
- C. Sleep records
- D. Pentylenetertrazol

1457.The quality of EEG may vary widely with :
- A. Skill of the technician and the quality of apparatus
- B. Number and location of the electrodes
- C. Length of the record
- D. All of the above

1458.In which of the situation, CSF pressure is maximum:
- A. Infants on reclining
- B. Children on reclining
- C. Adults reclining
- D. Any age while sitting

1459.All of the following diseases may cause fasciculations except:
- A. Poliomyelitis
- B. Werdnig Hoffmann's disease
- C. Cervical spondylosis
- D. None of the above

Ans.: 1449. D   1450. D   1451. C   1452. C   1453. A   1454. C
1455. C   1456. D   1457. D   1458. D   1459. D

1460. In which of the following diseasee pupils accommodate but do not react to light:

A. Head injury
B. Pontine hamorrhage
C. Tabes dorsalis
D. III neve palsy

1461. In which, of the following conditions, a patent will be confused, agitated with abnormally large pupil:

A. Barbiturate poisoining
B. Heroin overdose
C. Dhatura poisoning
D. Pontine hemorrhage

1462. A 66 year-old man has seizures that begin with clonic activity of the left hand and spread to the left arm, then fact and then leg. Subscquently, he has transient paresis of the left arm. The lesion is probably located in the :

A. Right lateral cerebral cortex
B. Left lateral cerebral cortex
C. Midline
D. Anywhere

1463. A 55-year-old woman developed moderate interscapular backpain, paresis with heperflexia, loss of sensation just below umbilicus and incontinence. The lesion is in the cord segment:

A. T_8
B. T_{10}
C. T_{12}
D. Cauda equina

1464. Subacute (in weeks) paralytic muscular disorders include all of the following except:

A. Botulism
B. Acute polyneuritis
C. Alcoholic myopathy
D. None of the above

1465. A 56-year-old man thought to have depression was then found to have right optic atrophy papilloedema of the left optic nerve and left hemiparesis. The lesion is in :

A. Left frontal lobe
B. Right frontal lobe
C. Right cerebellum
D. Left occipital lobe

1466. Pseudobulbar palsy in contrast to bulbar palsy is characterized by all the following except :

A. Emotional and intellectual impairment
B. Hyperactive Jaw jerk
C. Voluntary movement of palate
D. Babinski sign usually present

1467. A middle-aged complains of impotence. He had been in excellent health except hypertension. On examination, he had orthostatic hypotension and lightheadedness, but the neurologic examination is otherwise normal. On mental state examination, he had mild depression. The most probable cause is :

Ans.: 1460. C   1461. C   1462. A   1463. B   1464. D   1465. B
1466. C   1467. D

A. Multiple sclerosis   B. Renal artery stenosis
C. Hypertension   D. Antihypertensives

1468. **All of the following signs may be present in bilateral or diffuse cerebral disease except:**
A. Ideomotor apraxia   B. Perseverations
C. Pseudobulbar palsy   D. Gerstmann's syndrome

1469. **Which of the following is usually a sign of dominant hemisphere lesion:**
A. Constructional apraxia   B. Somatopagnosia
C. Gerstmann's syndrome   D. Anosognosia

1470. **Lumbar puncture may be useful in all of the following except:**
A. Subarachnoid hemorrhage   B. Cryptococcal meningitis
C. Tabes dorsalis   D. Hepatic encephalopathy

1471. **"Triphasic waves" in EEG may be found in:**
A. Wilson's disease   B. Porphyria
C. Water intoxication   D. Hepatic encephalopathy

1472. **"Periodic complexes" in EEG may be found in:**
A. SSPE   B. Creutzfeldt-Jacob's disease
C. Herpes encephalitis   D. All of the above

1473. **EEG (Electroencephalography) is of diagnostic assistance in all of the following except :**
A. Frontal lobe tumours
B. Sphenoid wing meningioma
C. Myoclonic epilepsy
D. Park inson-dementia complex

1474. **The conditions which may predispose to normal pressure hydrocephalus except:**
A. Subarachnoid hemorrhage   B. Chronic meningitis
C. Encephalitis   D. None of the above

1475. **The test used to diagnose normal pressure hydrocephalus is :**
A. CT Scan
B. Pneumoencephalography or cisternography with radioactive substances
C. CSF infusion test
D. All of the above

1476. **The treatment of choice in dementia caused by normal pressure hydrocephalus is:**
A. Antibiotics   B. Nootropics
C. Ventricular–peritoncal shunt D. None of the above

Ans.: 1468. D   1469. C   1470. D   1471. D   1472. D   1473. D
1474. D   1475. D   1476. C

1477. All of the following are the causes of dementia except :
  A. Tuberous sclerosis
  B. Bromism
  C. Normal or low pressure hydrocephalus
  D. Arsenic poisioning

1478. Myxedema madness most commonly resembles:
  A. Delirium   B. Dementia
  C. Depression   D. Mania

1479. Wernicke's encephalopathy has all of the following features except:
  A. Ocular paresis   B. Ataxia
  C. Peripheral neuropathy   D. Seizures

1480. An illness presenting as rapidly developing dementia, pyramidal and extrapyramidal motor findings and myoclonus:
  A. Wilson's disease
  B. Normal pressure hydrocephalus
  C. Porphyria
  D. Creutzfeldt-Jacob's disease

1481. A Condition with adenoma of face seizures and dementia, usually beginning in childhood:
  A. Obstructive hydrocephalus   B. Bromism
  C. Porphyria   D. Tuberous sclerosis

1482. Asterixis and confusion are typical of :
  A. SSPE   B. Arsenic poisoning
  C. Hepatic encephalopathy   D. Wilson's disease

1483. Watson – Schwartz test is used in the diagnosis of :
  A. Wilson's disease   B. Bromism
  C. Porphyria   D. Wernicke's encephalopathy

1484. Which of the following tests, may be useful in the diagnosis of Creutzfeldt – Joacob's disease :
  A. CSF   B. CT scan
  C. EEG   D. Serum electrolytes

1485. The term " Knife blade atrophy" or "Walnut brain" are applied to atrophy of gyri in:
  A. Alzheimer's disease   B. Pick's disease
  C. Huntington's chorea   D. Neurosphilis

1486. Which of the following is least commonly involved in Huntingtons chorea:
  A. Caudate nucleus   B. Putamen
  C. Frontal lobes   D. Parietal lobes

Ans.: 1477. D   1478. A   1479. D   1480. D   1481. D   1482. C
       1483. C   1484. C   1485. B   1486. D

1487.In Huntington's chorea, there is marked decrease in enzyme:
A. DOPA decarboxylase
B. Acetylcholinesterase
C. Choline synthetase
D. Glutamic acid decarloxylase

1488.In Hallervorden – Spatz disease, there is involvement of :
A. Caudate nucleus
B. Putament
C. Globus pallidus and substantia nigra
D. Frontal lobes

1489.In Lafora's disease, the most frequent epilepsy is :
A. Petitmal
B. Complex partial
C. Myoclonic jerks
D. Akinetic

1490.Lafora's disease is characterized by all except:
A. Autosomal recessive
B. Dementia
C. Begins in adulthood
D. Lafora bodies is the diagnostic morphologic feature

1491.The condition closely related to multiple sclerosis is :
A. Diffuse sclerosis of schilder
B. Adrenoleukodystrophy
C. Neuromyelitis optica or Devic's syndrome
D. All of the above

1492.Guillain Barre syndrome is most often associated with:
A. An acute viral or mycoplasma infection
B. Surgery
C. Pregnancy
D. Lymphoma or carcinoma

1493.Following autonomic signs are seen in Guillain Barre syndrome except:
A. Sinus bradycardia
B. Orthostatic hypotension
C. Excessive sweating
D. Hypertension

1494.Lewy bodies are seen in:
A. Alzheimer's disease
B. Pick's disease
C. Idiopathic Parkinsonism
D. SSPE

1495.Pathological finding which is specific for Alzheimer's disease is:
A. Senile plaques
B. Nurofibrillary tangles
C. Granulovacuolar degeneration
D. None of the above

Ans.: 1487. D   1488. C   1489. C   1490. C   1491. D   1492. A
1493. C   1494. C   1495. D

1496. All of the following are correct about **Pick bodies except:**
  A. Oval on shape
  B. Cytoplasmic
  C. Filamentous
  D. Found in blood vessels

1497. Which of the following is **not an autosomal recessive disease:**
  A. Tay-Sachs disease
  B. Friendreich's ataxia
  C. Werdnig – Hoffmann's disease
  D. Duchenne's muscular dystrophy

1498. All of the following are **nongenetic diseases except:**
  A. Syndenham's chorea
  B. Migraine
  C. Creutzfeldt-Jacob's disease
  D. Hypokalemic periodic paralysis

1499. Match the following:

	Sign		*Lesion*
I.	Anosognosia	i.	Basal ganglia
II.	Chorea	ii.	Right parietal lobe
III.	Aphasia	iii.	Left parietal lobe
IV.	Gerstmann's syndrome	iv.	Left temporal lobe

  A. I (i), II (ii), III (iii), IV (iv)
  B. I (ii), II (i), III (iii), IV (iv)
  C. I (ii), II (i), III (iv), IV (iii)
  D. I (iii), II (i), III (ii), IV (iv)

1500. Find out the **wrong match :**

Tendon reflex	*Nerve root*
A. Biceps	$C_5 - C_6$
B. Brachioradialis	$C_4 - C_5$
C. Ankle (gastronemius)	$L_5 - S_1$
D. Knee (quadriceps)	$L_2 - 4$

1501. Which of the following disease is associated with **faciculations :**
  A. Guillain – Barre syndrome
  B. Spinal cord compression
  C. Porphyria
  D. Psychogenic stress

1502. A **Young woman presents with confusion and hallucinations, flaccid paresis and abdominal pain. Her urine is red. She has :**
  A. Tabes dorsalis
  B. Hemolytic anemia
  C. Acute intermittent porphyria
  D. Hysteria

1503. Which of the following diseases is associated with **disturbances in the conduction apparatus:**
  A. Hypokalemic periodic paralysis
  B. Pernicious anemia
  C. Botulism
  D. All of the above

Ans.: 1496. D   1497. D   1498. D   1499. C   1500. B   1501. D
  1502. C   1503. D

1504.Electromyogrophy shows bizarre high frequency discharges without any waxing or waning. This is characteristic of :
- A. Normokalemic periodic paralysis
- B. Myasthenia gravis
- C. Myotonia congenital ·
- D. Grave's disease

1505.Episodic skeletal muscle Paralysis is characteristic of :
- A. Erb's and Leyden Moebius disease
- B. Subacute combined degeneration
- C. Myasthenia gravis
- D. Myotonia congenital

1506.Which of the following diseases probably has viral actiology:
- A. Subacute sclerosing panencephalitis (SSPE)
- B. Guillain Barre syndrome
- C. Creutzfeldt – Jacob's disease
- D. All of the above

1507.In Reye syndrome, brain edema is mainly:
- A. Vasogenic
- B. Cytotoxic
- C. Interstitial
- D. All of the above

1508.Following are the causes of vasogenic brain edema except:
- A. Brain tumor
- B. Ischemia
- C. Purulent meningitis
- D. Water intoxication

1509.Following are the causes of cytotoxic brain edema except:
- A. Disequilibrium syndromes
- B. Water intoxication
- C. Ischemia and purulent meningitis
- D. None of the above

1510.Following may produce interstitial type of brain edema except:
- A. Obstructive hydrocephalus
- B. Purulent meningitis
- C. Pseudotumor cerebri
- D. Lead encephalopathy

1511.Steroids are most beneficial in which type of brain edema:
- A. Vasogenic
- B. Cytotoxic
- C. Interstitial
- D. All of the above

1512.Osmotherapy is most useful in which type of brain edema:
- A. Vasogenic
- B. Cytotoxic
- C. Interstial
- D. All of the above

1513.Acetazolamide has main usefulness in brain edema:
- A. Vasogenic
- B. Cytotoxic
- C. Interstitial
- E. All of the above

Ans.: 1504. D   1505. C   1506. D   1507. B   1508. D   1509. D
1510. D   1511. A   1512. B   1513. C

1514.Which of the following type of fluid should be avoided in brain edema:

A. Normal saline
B. 5% glucose in saline
C. 5% dextrose in saline
D. 5% glucose

1515.The following drugs may produce benign intractranial hypertension except:

A. Lithium carbonate
B. Nalidixic acid
C. Amiodarone or chlordecone
D. Spironolactone

1516.In adults, which of the following is an uncommon symptom of normal pressure hydrocephalus (NPH) :

A. Altered consciousness
B. Papilloedema
C. Seizures
D. Ataxia and corticospinal tract signs

1517.The following drugs have been tried in the treatment of normal pressure hydrocephalus, with or without lumbar puncture, except:

A. Acetazolamide
B. Furosemide
C. Isosorbide
D. Nimodipine

1518.An increase in the blood volume of the brain caused by obstruction of the cerebral veins and venous sinuses or by arterial vasodilatation such as that caused by hepercapnial is know as :

A. Brain edema
B. Brain engorgement
C. Hydrocephalus
D. Any of the above

1519.Following are the main 3 types of brain edema except :

A. Vagogenic
B. Cellular (cytotoxic)
C. Exvacuo
D. Interstitial (hydrocephalus)

1520.In which of the following types of brain edema, location of edma is the both gray and white matter" :

A. Vasogenic
B. Cellular
C. Interstitial
D. All of the above

1521.In Which type of brain edema, capillary permeability to large molecules (Insulin etc) is increases:

A. Vosogenic
B. Cellular
C. Interstitial
D. None of the above

1522.The origin of uncinate seizures is usually from :

A. Frontal lobe
B. Temporal lobe
C. Parieatal lobe
D. Occipital lobe

Ans.: 1514. D   1515. D   1516. C   1517. D   1518. B   1519. C
1520. B   1521. A   1522. B

1523. A 66-year-old man has the sudden, painless onset of paresis of the right upper and lower face, inability to abduct the right eye, and paresis of the left arm and leg. The lesion is most likely in the :

A. Midbrain
B. Medulla
C. Pons
D. Temporal lobe

1524. The Structures involved in the above case are :

A. VI, VII nerves on the left and cortiospinal tract on the right
B. VI, VII nerves on the right and corticospinal tract on the left.
C. VI, VII nerves and corticospinal tract on the right side.
D. VI, VII nerves and corticospinal tract on the left side

1525. A 24-year-old man has the subacute onset of complete loss of vision in the left eye incoordination of the right hand the lesion is in:

A. Left optic nerve and cerebellar system
B. Right optic nerve and cerebellar system
C. Left optic nerve and right cerebellum
D. Right option nerve and left cerebellum

1526. All of the following are the signs may be present when either of the hemisphere is involbe except:

A. Hemiparesis
B. Hemisensory loss
C. Homonymous hemianopsis
D. Aphasia

1527. Which of the following signs is usually present only when both hemispheres are involved :

A. Focal seizures
B. Constructional apraxia
C. Aphasias
D. Pesudodollr palsy

1528. Which of the following sign indicates a lesion in the basal ganglia:

A. Hemiparesis
B. Contructional apraxia
C. Rigidity
D. Alternating hyalgesia

1529. Hemiballismus is usually the result of a small cerebrovascular infarction of the :

A. Basal ganglia
B. Pons
C. Ipsilateral corpus of lyusii
D. Contralateral corpus of Lyusil

1530. The influence of smell on peychosexual behaviour is mediated by cranial nerve:

A. I
B. II
C. V
D. X

1531. For testing the function of first cranial some irritative volatite substances such as ammonia is most suitable as they stimulate the sensory nerve endings of:

A. V
B. VII
C. XI
D. None of the above

Ans.: 1523. C   1524. C   1525. C   1526. D   1527. D   1528. C
1529. D   1530. A   1531. A

1532.Anosmia may be present in :
   A. Olfactory groove meningioma
   B. Hysteria
   C. Frontal lobe tumor
   D. All of the above

1533.All of the following are the causes of coma along with miosis except:
   A. Heroin intake
   B. Pontine interactive
   C. Internal capsule infarction
   D. Barbiturate intoxication

1534.Which of the following drugs may cause myopathy:
   A. Chlorpromazine   B. Nitrofurantion
   C. Hydrochlorthiazide   D. INH

1535.Which of the following drug does not lead to neuropathy:
   A. INH   B. Nitrofurantoin
   C. Phenytoin   D. Prednisolone

1536.A person had a left superior homonymous quadrantanopsia. The probable cause can be all of the following except:
   A. Right temporal lobe lesion
   B. Superior occipital lobe lesion
   C. Inferior occipital lobe lesion
   D. Optic tract lesion occasionally

1537.In which of the following situation, the presence of frontal release reflexes will indicate a structural brain lesion:
   A. In infants
   B. In adults
   C. A patient with schizophsenia
   D. The presence of unilateral signs only

1538.All are features of Resfum disease except:
   A. Neurogenic weakness
   B. Stocking-and gloves sensory loss
   C. Exaggerated reflexes
   D. Thickened nerves

1539.Which is primary site of damage in Wilson's disease, Huntington's chorea and choreiform cerebral palsy:
   A. Cerebrum   B. Midbrain
   C. Basal ganglia   D. Corticospinal tract

Ans.: 1532. D   1533. C   1534. C   1535. D   1536. B   1537. D
      1538. C   1539. C

1540. A 56-year-old man with mild dementia has absent reflexes, loss of position and vibration sensation and ataxia. The nonstructural disease for these symptoms may be:
   A. Pernicious anemia   B. Tabes dorsalis
   C. Heavy metal intoxication   D. All of the above

1541. An injury to right third cranial nerve will result in all the following except:
   A. Right upper lid paresis
   B. Abduction in right eye
   C. Dilated pupil
   D. Diplopia on looking down and medially

1542. A young lady complained of vertigo, nausea vomiting and left-sided tinnitus. On examination he had nystagmus to the right but no papilloedema. The most likely cause can be :
   A. Left-sided inner ear disease B. Right-sided inner ear disease
   C. Right cerebellar lesion   D. Left cerebellar lesion

1543. All of the following neurologic diseases have autosomal dominant transmission except:
   A. Charcot-Marie-Tooth disease
   B. Ataxia telangietasia
   C. Hypokalemic periodic paralysis
   D. Dystonia musculorium deformans

1544. The half-life of the following antiepileptic is shorter in children:
   A. Phenobarbitone   B. Ethosuximide
   C. Trimethadione   D. All of the above

1545. Which of the following is maximum protein bound:
   A. Phenytoin   B. Carbamazepine
   C. Phenobarbitone   D. Clonazapam

1546. Diplopia without nystagmus is in important side effect of :
   A. Phenytoin   B. Phenobarbitone
   C. Carbamazepine   D. Sodium Valproate

1547. Bone marrow suppression is caused by all except:
   A. Carbamazepine   B. Ethosuximide
   C. Sodium valproate   D. Phenobarbitone

1548. The Second most frequent cause of infectious dementia is :
   A. AIDS   B. Creutzfeldt-Jakob disease
   C. Neurosyphilis   D. Fungal meningitis

Ans.: 1540. D   1541. D   1542. A   1543. D   1544. D   1545. A
        1546. C   1547. D   1548. B

1549.In Biswanger dementia, all of the following pathological brain changes are seen except:
   A. Amyloid like pigment   B. Lacunae
   C. Cysts   D. Demyelination

1550.Biwanger dementia occurs commonly in association with :
   A. Diabetes mellitus   B. Hypertension
   C. Hepatic encephalopathy   D. None of the above

1551.Match the following:

Antiepileptics	*Half life (approx.)*
I. Phenobarbitone	i. 6 – 13 days
II. Carbamazepine	ii. 24 hours
III. Phentoin	iii. 13-17 hours
IV. Trimethadione	iv. 90 hours

   A. I (iv), II (iii), III (i), IV (ii)   B. I (ii), II (iii), III (i), IV (iv)
   C. I (iii), II (ii), III (i), IV (iv)   D. I (iv), II (iii), III (ii), IV (i)

1552.Ataxia is a side effect of:
   A. Phenytoin   B. Carbamazepine
   C. Ethosuximide   D. All of the above

1553.Decreased libido is an important side effect of :
   A. Clonazepam   B. Trimethadione
   C. Phenytoin   D. Phenobarbitone

1554.Skin rash may be produced by :
   A. Phenytoin   B. Phenobarbitone
   C. Ethosuximide   D. All of the above

1555.Sodium valproate if given with Clonazepam may precipitate:
   A. Nephrotic syndrome   B. Cerebellar syndrome
   C. Absences status   D. Mania

1556.Sodium valproate increases the blood levels of all of the following except:
   A. Carbamazepine   B. Phenobarbitone
   C. Methsuximide   D. None of the above

1557.Following may cause preudotumor cerebri except:
   A. Empty-sella syndrome
   B. Meningism with systemic bacterial or viral infections
   C. Lupus erythematous
   D. Dermatomyositis

1558.Following are the endocrinal causes of pseudotumor cerebri except
   A. Hyperadrenalism   B. Hyperthyroidism
   C. Addison's disease   D. Hypoparathyroidism

Ans.: 1549. A   1550. B   1551. D   1552. D   1553. D   1554. D
      1555. C   1556. B   1557. D   1558. B

1559. **All of the following findings throw doubt on the diagnosis of benign intracranial hypertension (Pseudotumor cerebri) except:**
    A. LSF protein greater than 50 mg%
    B. Decreased CSF glucose
    C. Increased cell count
    D. Normal EEG

1560. **Pseudopapilloedema (a developmental anomaly of the fundus) is characterized by:**
    A. Elevation of the optic disc   B. Exudates
    C. Hemorrhages   D. Decreased visual acquity

1561. **In perimetry, pseudopapilloedema may show:**
    A. Bjerrum's scotoma   B. Siedel's sign
    C. Enlargement of blind spot   D. Annular scotoma

1562. **Optic neuritis differs from pseudopapilloedema by all of the following except:**
    A. Visual loss
    B. Normal CSF pressure
    C. Changing appearance of fundus
    D. None of the above

1563. **In females, following may produce benign intracranial hypertension (BIH) except :**
    A. Menarche   B. Pregnance
    C. Female sex hormones   D. Menopause

1564. **In pseudopapilloedema, which of the following is the most diagnostic sign in fundus:**
    A. Exudates
    B. Hemorrhages
    C. Elevation of optic disc
    D. Unchanging appearnce of fundus on repeated examinations

1565. **In benign intracranial hypertension, which of the following symptoms may be an indication for shunt operation:**
    A. Intractable headache   B. Diplopia
    C. Ataxia   D. None of the above

1566. **Following may be the presentations of intracranial hypotension except:**
    A. Severe headache aggravated by erect posture
    B. Unilateral or bilateral VI nerve palsy
    C. Nausea and dizziness
    D. Creeping anaesthesias

Ans.: 1559. D   1560. A   1561. C   1562. D   1563. D   1564. D
       1565. A   1566. D

1567.The most common cause of intracranial hypotension is :
A. Head injury
B. Bacterial meningitis
C. CSF rhinorrhea in brain tumor
D. Lumbar puncture

1568.Treatment of intracranial hypotension includes:
A. Bed rest
B. Injection of patient's own blood (10 – 20 ml) into the epitural space ("blood patch")
C. Surgical closure of defect in dura
D. All of the above

1569.All of the following are the frontal lobe release signs except:
A. Snout reflex   B. Palmomental reflex
C. Gag reflex   D. Grasp reflex

1570.A 46-year-old woman, who was entirely well previously, had the sudden onset of jargon speech and hysteria. She was admitted to the psychiatric ward with a diagnosis of undifferentiated, schizophrenia. The physical examination reveals only a right Babinski sign and an unequivocal right hemiparesis. The most probable diagnosis is :
A. Malingering   B. Drug induced parkinsonism
C. Left temporoparietal lesion   D. Right temporaparietal lesion

1571.Which of the following conditions may cause asymptomatic miosis:
A. Tomalol   B. Eserine
C. Old age   D. Pons lesion

1572.A patient on looking towards the left has diplopia. The cause can be:
A. Left VI nerve lesion
B. Left III cranial nerve lesion
C. Right-side internuclear ophthalmoplegia
D. All of the above

1573.A young man sustained major head trauma. After wards, he had intellectual and personality changes and reported that food tasted differently. The probably cause can be:
A. Lesion in frontal lobe   B. Lesion in temporal lobe
C. Lesion in Mid brain   D. Hysteria

1574.All of the following are the signs of occlusion of unilateral internal carotid artery except:
A. Hemiparesis   B. Homonymous hemianopsia
C. Dementia   D. Aphasia

Ans.: 1567. D   1568. D   1569. C   1570. C   1571. D   1572. B
1573. A   1574. C

1575. A 40-year-old patient had hemiparesis, hemisensory loss and homonymous hemianopsia and does not describe any deficit or rather denies it. He is most likely have lesion in the :
A. Right hemisphere
B. Left hemisphere
C. Both hemispheres
D. Basal ganglia

1576. A 16-year-old married lady complains of right eye blindness, right hemiparesis, and right hemisensory loss. Papillary and tendon reflexes are normal and there is no other neurological deficit. Hoover's sign is present. The lesion is most likely explained by :
A. Left internal carotid artery occlusion
B. Right internal carotid artery occlusion
C. Lesion in internal capsule
D. Hysteria

1577. The disease associated with movement disorders and dementia include:
A. Wilson's disease
B. Syndenham's chorea
C. Huntington chorea
D. Creutzfeldt-Jabok's disease

1578. In Alzheimer's disease, the most frequent signs are of :
A. Frontal lobe
B. Parietal lobe
C. Occipital lobe
D. Basal ganglia

1579. Trophic ulcers in fingers are often associated with the following conditions except:
A. Leprosy
B. Syringomyelia
C. Cervical disc prolapse
D. Subacute combined debeneration

1580. Raised intracranial pressure is suspected if there is :
A. Episodic throbbing unilateral headache
B. Unilateral ptosis
C. VI nerve palsy
D. Neck stiffness

1581. All of the following are the signs indicating a lesion in the non-dominant hemisphere:
A. Anosognosia
B. Somatopagnosia
C. Aphasia
D. Constructional apraxia

1582. Which of the following sign indicates brain stem lesion:
A. Bulbar palsy
B. Pseudobulbar palsy
C. Anosognosia
D. Seizures

Ans.: 1575. A 1576. D 1577. B 1578. A 1579. D 1580. C
1581. C 1582. A

1583. Which of the following visual fielf defect indicates a lesion in the temporal lobe:
A. Homonymous hemianopsia   B. Superior quadrantanopsia
C. Bitemporal hemianopia   D. Binasal hemianopia

1584. A. patient had right-sided hemiparesis and left-sided facial paralysis of lower motor neurone type. The lesion is in the :
A. Midbrain   B. Right side of pons
C. Left side of pons   D. Lateral medulla

1585. Which of the following is a sign of lesion in the cerebellum:
A. Bradykinesia   B. Sexual dysfunction
C. Somatopagnosia   D. Dysmetria

1586. Which of the following signs is usually present when both hemispheres are involved:
A. Homonymous hemianopia   B. Hemiparesis
C. Jacksonian epilepsy   D. Dementia

1587. A patient with amyotrophic lateral sclerosis will have all of the following except:
A. Decreased sensation
B. Fasciculations
C. Hyper reflexic deep tendon reflexes
D. Babinski sign

1588. Unlike cerebellar dysfunction, movement disorders from basal ganglia disease have all of the following features except:
A. Prominent during rest
B. Absent during sleep
C. Diminished by concentration and intentional movements
D. None of the above

1589. Treatment of choice of Rabbit syndrome (Orofacial dyskinesias produced by long term intake of neuroleptics) is :
A. Choline   B. Trihexyphenidyl
C. Lithium   D. Propranolol

1590. The drug which may be useful in the treatment of tardive dyskinesia is :
A. Lecithin   B. Amoxapine
C. Biperidine   D. Atenolol

1591. Hyperosmolar solutions (such as Mannitol) are useful in the treatment of acute stroke only if used in:
A. 1st week   B. 2nd week
C. 3rd week   D. 4th week

Ans.: 1583. B   1584. C   1585. D   1586. D   1587. A   1588. D
1589. B   1590. A   1591. A

1592.Administration of mannitol may cause a transient rise in :
A. Cerebral thrombosis
B. Blood urea
C. SGOT
D. Uric acid

1593.Mannitol is used in acute stroke as:
A. 5%
B. 9%
C. 20%
D. 40%

1594.Glycerol is used in acute stroke as :
A. 1%
B. 5%
C. 10%
D. 20%

1595.Does of glycerol per kg body weight in the treatment of stroke is :
A. 01. – 0.5 g
B. 0.5 – 1.0 g
C. 1.0 – 1.5 g
D. 1.5 – 2.5 g

1596.Rebound increase in intracranial pressure is least reported with the use of :
A. Mannitol
B. Glycerol
C. Corticosteroids
D. None of the above

1597.If corticosteroids are used for cerebral decongestions, which of the following should be monitored:
A. Blood pressure
B. ECG
C. Serum potassium
D. Urine output

1598.The actions of low molecular weight dextrin include:
A. Increase in cerebral blood flow
B. Preventing platelet aggregation
C. Increase in cardiac output and decrease in blood viscosity
D. All of the above

1599.The use of low molecular weight dextran is contraindicated in patients with:
A. Liver disease
B. Bronchial asthma
C. Incipient left heart failure
D. All of the above

1600.Xantinol nicotinate (complamina) is best used as a vasodilator in the treatment of crebral infarction by route:
A. Orally
B. Intravenously
C. Intramusccularly
D. Subcutaneously

1601.The risk of intravenous use of Xantinol nicotinate is :
A. Arrythmias
B. Hypotension
C. Seizures
D. Idiosyncrative reactions

1602.The use of Xantinol nicotinate is associated with:
A. Peptic ulcer
B. Ulcerative colitis
C. Jaundice
D. Inefficacy

Ans.: 1592. B   1593. C   1594. C   1595. C   1596. B   1597. C
1598. D   1599. C   1600. C   1601. B   1602. D

1603. Which of the following drug has not been found useful in the treatment of cerebral infarction:
  - A. Xanthinol nicotinate
  - B. Oxygen therapy
  - C. Streptokinase
  - D. Frusemide

1604. All of the following phenomena may be seen both in REM and NREM sleep except:
  - A. Restless leg syndrome
  - B. Head banging
  - C. Seizures
  - D. Complex motor activity

1605. Which of the following may be seen in both REM and NREM sleep:
  - A. Bruxism
  - B. Complex intellectual activity
  - C. Night terrors
  - D. Nightmares

1606. Which of the following statement is incorrect:
  - A. Sleep apnea is a disorder only of adults
  - B. Sleep apnea is sometimes associated with narcolepsy
  - C. Sleep apnea leads to cardiovascular disturbances as well as excessive day time sleepiness
  - D. Hypnopompic refers to phenomena that occur on awakening and hypnogogic refers to phenomena that occur on falling asleep.

1607. Match the following:

Tremor	*Test*
I. Action or postural tremor	i. Finger-nose test
II. Cerebellar tremor	ii. Causing psychologic stress
III. Parkinsonian tremor	iii. Extending arms and hands
IV. Anxiety-induced tremor	iv. Any of the above

  - A. I (i), II (ii), III (iii), IV (iv)
  - B. I (ii), II (iii), III (iv), IV (i)
  - C. I (iii), II (i), III (ii), IV (iv)
  - D. I (iv), II (iii), III (ii), IV (i)

1608. Which of the following is not true about Oculogyric crisis:
  - A. Occurs in the course of treatment
  - B. Improves when the modication is reduced
  - C. Occurs more commonly with older patients
  - D. Responds to anticholinergic medications

1609. In brain tumors (irrespective of site), the most frequent symptom is:
  - A. Nausea and vomiting
  - B. Seizures
  - C. Personality deterioration
  - D. Intellectual deterioration

1610. Most reliable investigation in brain tumour is:
  - A. EEG
  - B. Isotop scanning
  - C. Cerebral angiography
  - D. CT Scanning

Ans.: 1603. C   1604. D   1605. A   1606. A   1607. C   1608. C
1609. B   1610. D

1611. The sensory part of Gag Reflex is mediated by _____ cranial nerve :
A. VII   B. IX
C. X   D. XII

1612. Root value of finger jerk is :
A. C5   B. C6
C. C7   D. C8

1613. Acetylcholine, an excitatory neurotransmitter has clinical relevance in following except :
A. Motion sickness   B. Huntington's chorea
C. Bladder control   D. Migraine

1614. Which of the following group of neurotransmitters may act as both excitatory and inhibitory :
A. GABA   B. Histamine
C. Neuropeptides   D. Purines

1615. Normally CSF has about more than ——% of blood level of glucose :
A. 40   B. 50
C. 60   D. 80

1616. Protein content is CSF is normal/increased in following except :
A. Subarachnoid haemorrhage
B. AC. bacterial meningitis
C. Viral meningitis
D. TBM

1617. In following conditions, CSF is microbiologically sterile except :
A. Normal   B. SAH
C. Multiple sclerosis   D. Ac. bacterial meningitis

1618. Following are features of headache of raised intracranial pressure except :
A. Associated with morning vomiting
B. Worse with cough and straining
C. Worse bending forward
D. Worse in evening

1619. The risk of thromboembolic stroke is increased in Migraine in following except :
A. Migraine with aura
B. Migraine without aura
C. Patient on oral contraceptive
D. Patient on Ergot compound

1620. Following is to be avoided in patient on treatment for migraine:
A. Triptans   B. Codeine containing analgesics
C. Propranolol   D. Pizotifen

Ans: 1611. B   1612. D   1613. D   1614. C   1615. C   1616. C
      1617. D   1618. D   1619. D   1620. B

1621. **Following is incorrect about migrainous neuralgia (cluster headache):**
   A. 10-50 times less common than migraine
   B. 5.1 predominant in females
   C. Onset is usually in third decade
   D. Subcutaneous injection of sumitriptan for acute attack·

1622. **Trigeminal neurolgia typically occurs in patients older than _____ years :**
   A. 35   B. 40
   C. 45   D. 55

1623. **In those under 65 years, the commonest cause of dizziness is:**
   A. Head injury   B. Anxiety
   C. Epilepsy   D. Drugs

1624. **In versive seizures, the force eyes show :**
   A. Forced deviation to same side
   B. Forced deviation to opposite side
   C. Front eye field may be involved
   D. Eyes remain in center

1625. **Which of the following seizures have age of onset in 4-8 years:**
   A. Juvenile absence epilepsy
   B. Juvenile myoclonic epilepsy
   C. Childhood absence seizures
   D. GTCS

1626. **Following drugs have been implicated in producing secondary generalized epilepsy except :**
   A. Ciclosporin   B. Lidocaine
   C. Penicillin   D. Azithromycin

1627. **Following are genetic causes of partial seizures except :**
   A. Down's syndrome   B. Tuberous sclerosis
   C. Neurofibromatosis   D. Von-Hippel-Lindau disease

1628. **Acetazolamide is used in following seizure types except :**
   A. Primary and secondary GICS
   B. Myoclonus
   C. Absences
   D. Partial

1629. **Following is the therapeutic range (μmol/L) of oxcarbazepine:**
   A. 10-40   B. 40-80
   C. 80-120   D. 120-160

1630. **Gapapentin interacts with :**
   A. Antacids   B. Antimalarials
   C. Aspirin   D. Antidepressants

Ans: **1621. B**   **1622. D**   **1623. B**   **1624. B**   **1625. A**   **1626. D**
1627. A   **1628. B**   **1629. B**   **1630. A**

SECTION - III

MCQ's asked in previous
PG & Postdoctoral Entrance Exams

MCQ's asked in previous
PG & Postdoctoral Entrance Exams

1. Pseudopapilloedema differs from papilloedema by the absence of the following except :
   PGI 1986
   A. Haemorrhage
   B. Engorgement of retinal vessels
   C. Decreased visual acquity
   D. All of the above

2. By increase in intracranial pressure the fluids pressure is primarly increased in :
   AMU 1985
   A. Subdural sheath around optic nerve
   B. Subarachnoid space around optic nerve
   C. Optic nerve itself
   D. All of the above

3. By increase in intracranial pressure, papilloedema is due to:
   AIIMS 1986
   A. Compression of retinal vein
   B. Compression of retinal artery
   C. Dilatation of retinal vein
   D. Not known

4. Vomiting is prominent in :
   AMU 1985
   A. Frontal lobe meningioma   B. Parietal lobe glioma
   C. Frontal lobe atrophy   D. Focal lesion in IV ventricle

5. Patients with thymoma and myasthenia graivis should receive anticholinesterase up to ——before surgery:
   A. 3-4 hr   B. 6-8 hr   AMU 1987
   C. 12-24 hr   D. 24-48 hr
   E. 48-72 hr

6. Which of the following is the commonest bacterial cause of brain abscess after trauma involving multiple fractures of skull:
   A. Staphylococcus   B. Streptococcus   AMC 1987
   C. H.influenzae   D. Esch coli

7. Find the wrong match:
   AIIMS 1984
Age group	Cause of meningitis
A. Infant	Esch.coli
B. Below 4 years	H.influenzae
C. Adults	Meningococcus
D. None of the above	

Ans.: 1. D   2. B   3. A   4. D   5. B   6. A   7. D

8.   The bacteria most often associated with meningitis secondary to ear infection :
A. Meningococcus   B. Pneumococcus   **AIIMS 1985**
C. Esch.coli   D. Staphylococcus

9.   The commonest cause of secondaries in   brain is tumours of:
A. Lung   B. Breast   **UPSC 1988**
C. Larynx   D. Liver

10.   The following intracranial secondaries may cause bony changes except :
                                             Rohtak 1988
A. Hodgkin's lymphoma   B. Carcinoma stomach
C. Secondaries   D. Multiple myeloma

11.   The following bone tumour may cause dural deposted without causing bony changes :   **AIIMS 1986**
A. Hodgkin's lymphoma   B. Multiple myeloma
C. Secondaries   D. Fibrous dysplasia

12.   Sturge-Weber syndrome is characterized by all of the following except :   **AIIMS 1988**
A. Mental retardation
B. Visual disturbances
C. Angiomas of choroid and pia mater
D. Renal anomalies

13.   A dome shaped skull with a high fore head in the infant with slight hydrocephalus (olympian brow) is seen in: **AIIMS 1984**
A. Marasmus   B. Congenital syphilis
C. Rickets   D. ArnoldChiari syndrome

14.   In acute head injury, immediate operation is indicated by :
A. Bloody spinal fluid on lumbar tap   **UPSC 1985**
B. Unilateral fixed dilated pupil
C. Rapidly changing neurological
D. Generalised convulsive seizures   **(Bailey-552)**

15.   Patient with a history of fall, present a week later with headache and progressive neurological deterioration. The diagnosis is:
A. Acute subdural hemorrhage   **AI 1989**
B. Extradural hemorrhage
C. Chronic subdural hemorrhage
D. Fracture skull   **(Bailey-395)**

Ans.: 8. B   9. A   10. A   11. A   12. D   13. B   14. C   15. C

16. In a patient with a moderately severe head injury, the development of the following sign suggests an acute increase in the intracranial tension :
    A. A deterioration in the level of consciousness   **Manipal 1994**
    B. A decrease in the arterial blood pressure
    C. An increase in the heart rate
    D. An increase in the rate of respiration

17. True about blood picture in Eosinophilic granuloma is:
    A. ↑ Eosinophilis   **Delhi 1986**
    B. ↓ Eosinophils
    C. Normal
    D. Variable or normal eosinophils and not important for diagnosis

18. Wrist drop is seen in palsy of :
    A. Ulnar nerve   B. Radial nerve   **Delhi 1986**
    C. Median nerve   D A + B   **(Bailey-385)**

19. Feature of extradural haemorrage include all except :
    A. Severe hypotension   **AP 1987**
    B. Deteriorating consciousness
    C. Fixed dilated pupil on the same side
    D. Fracture line crossing the temporal bone   **(Bailey-390)**

20. The involvement of cortical lesions cause the following type of speech defect :
    A. Dysphasia   B. Dysarthria   **AIIMS 1987**
    C. Dysphonia   D. Disarticulation

21. A patient has meningomyelocele with paraplegia, true is:
    A. Can be improved by surgery   **AIIMS 1988**
    B. Can not be improved by surgery
    C. Will improve as he gain age
    D. None of the above   **(Schwartz-1924)**

22. Vomiting following head injury is: **Delhi 1983**
    A. A sign of recovery from cerebral contusion
    B. A sign of recovery from cerebral concussion
    C. A sign of recovery from cerebral laceration
    D. A sign of Omen   **(Bailey-390)**

23. Commonest calcified brain mass in children in suprasellar region is:
    A. Craniopharyngioma   B. Meningioma   **Delhi 1987**
    C. Tuberculoma   D. Medulloblastoma
        (Bailey-561)

Ans.: 16. A   17. C   18. B   19. A   20. A   21. A   22. D   23. A

24. **Commonest primary brain tumour is:**
    Delhi 1986; AIIMS 1986; PGI 1983, 86
    A. Astrocytoma   B. Glioblastoma
    C. Ependymoma   D. Meningioma   **(Bailey-559)**

25. **Commonest intramedullary spinal tumour is:**
    A. Chordoma   B. Meningioma   **Delhi 1986**
    C. Ependymoma   D. Oligodendroglioma
                                                   (Bailey-364)

26. **Bloody C.S.F. is seen in:**   **Delhi 1992**
    A. Sub-arachinoid haemorrhage
    B. Sub dural haemorrhage
    C. Extra Dural hematoma
    D. Cavernous sinus thromsosis   **(Bailey-563)**

27. **Seizures in a patient of age 50 year is suggestive of:**
    A. Trauma   B. Tumour   **Delhi 1992**
    C. Hysterical   D. C V A

28. **Subdural effusion is caused by :**
    Delhi 1992; AIIMS 1986, 92; AI 1993
    A. Haemophilus influenzae   B. Preumococcus
    C. Str.virdans   D. Staph.aureus

29. **Lumbar puncture is dangerous in :**   **Delhi 1984, 94**
    A. Spinal cord tumour   B. Renal failure
    C. Metabolic alkalosis   D. All of the above

30. **Complete functional loss in a peripheral nerve without anatomic disruption and sudden complete return of the function in 60-90 days is called :**
    A. Neuropraxis   B. Neurolysis   **UPSC 1984**
    C. Chronaxie   D. Demyelinization   **(Bailey-532)**

31. **For cystic lesions in brain, investigation of choice is:**
    A. Angiography   B. Ultrasound   **DNB 1990**
    C. CT scan   D. MRI

32. **Operative control of bleeding from wounds of the scalp is best achived by:**
    A. Diathermy to bleeding vessels   **Rohtak 1986**
    B. Eversion of galea aponeurotica
    C. Applying several forceps to the bleeding points
    D. Direct pressure applied to the skin   **(Bailey-548)**

Ans.: 24. A   25. C   26. A   27. D   28. A   29. A   30. A   31. D   32. B

33. **Subarachnoid haemorrhage is commonly due to:**
     UPSC 87 JIPMER 1992; WB 1995
     A. Hypertenon   B. Rupture of aneurysm
     C. Trauma   D. Stroke   **(Bailey-563)**

34. **The most common site of chondroma in vertebral column is:**
     JIMPER 1992
     A. Sacro-coccygeal   B. Cervico-Thoracic
     C. Thoracic   D. Thoraco lumber

35. **Management of extradural haemorrage is:**   **AIIMS 1992**
     A. Immediate evacuation   B. Evacuation after 24 hours
     C. Antibiotics   D. Observation   **(Bailey-552)**

36. **The first step in the management of head injury is:**
     AIIMS 1992; Rajasthan 1991, 1994
     A. Secure airway   B. I.V mannitol
     C. I.V Dexamethasone   D. Blood transfusion   **(Bailey-552)**

37. **Most common site of meningomyelocele is:**   **AIIMS 1992**
     A. Cervical spine   B. Lumbo sacral spine
     C. Thoracic spine   D. Skull   **(Schwartz-1924)**

38. **Most common tumour of posterior cranial fossa :**   PGI 1993
     A. Meningioma   B. Glioma
     C. Medullobastoma   D. Oligodendroglioma

39. **Glasgow coma scale includes all of the following except :**
     A. Eye opening   B. Verbal performance   **DNB 1991**
     C. Motor activity   D. Motor tone   **(Bailey-552)**

40. **Disorders of limbic system primarily affects:**   **AIIMS 1984**
     A. Recall of past memory
     B. Recognition of past memory
     C. Registration of new memory
     D. All of the above

41. **The commonest cause of focal temporal lobe epilepsy is:**
     A. Tumour   B. Idiopathic   **AIIMS 1984**
     C. Abscess   D. Tuberculoma

42. **The following are the blood gas changes associated with raised
     intracranial pressure:**   **AIIMS 1984, 1986**

$PaCO_2$	PaO_2
A. Increase	Increase
B. Increase	Decrease
C. Decrease	Increase
D. Decrease	Decrease

Ans.: 33. B   34. C   35. A   36. A   37. B   38. C   39. D   40. C   41. B

43.   The organism most often associated with subdural abscess:

Delhi 1984

A. Staphylococcus   B. Streptococcus
C. Pneumococcus   D. Meningococcus

44.   First symptom of papilloedema is :   Delhi 1985

A. Increased blind spot   B. Seeing moving objects
C. Gastric upset   D. Lacrimation

45.   Comonest site of meningocele: AI 1989

A. Lumbosacral   B. Occipital
C. Frontal   D. Thoracic   (Bailey-363)

46.   In chornic subdural haematoma, not seen is :

PGI 1980; AMC 1985

A. Always preceded with history of unconciousness
B. Stroke
C. Headache
D. Paraplegia or altered sensorium   (Bailey-551)

47.   Spinal shock due to contusion usually recovers in :

AIIMS 1983

A. 1-2 days   B. 5-7days
C. 7-14 days   D. 14-21 days

48.   Angular vein infection commonly causes thrombosis of —— sinus:

AIIMS 1982

A. Cavernous   B. Sphenoidal
C. Petrosal   D. Sigmoid

**49.   Infants and children irradiated for the treatment of an enlargement
of the thymus are at an increased risk of :**

Manipal 1995

A. Carcinoma of the thyroid   B. Leukaemia
C. Malignant thymoma   D. Mediastinal germinoma

50.   The rare of growth in nerves after peripheral nerve suturing is:

JIPMER 1986 AIIMS 1986, 87

A. 1 mm/day   B. 1.5 mm/ day
C. 2 mm/ day   D. 3 mm/day

51.   Most persisting symptom of head injury is: DNB 1990

A. Amnesia   B. Fits
C. Anosmia   D. Headache

52.   Cauda equina lesion may produce :   PGI 1983

A. Brown Sequard lesion   B. Quadriplegia
C. Any of the above   D. None of the above

Ans.: 42. B   43. B   44. A   45. A   46. A   47. A   48. A   49. A   50. A
      51. D   52. C.

53. In children,intracranial tumours are more common in:

PGI 1984, 1985

A. Anterior fossa   B. Middle fossa
C. Posterior fossa   D. Equal incidences in all fossa

54. Musculo skeletal abnormality in neurofibromatosis:

AMC 1983; PGI 1988

A. Hypertrophy of limb   B. Scoliosis
C. Cafe au lait spots   D. Pseudo arthrosis
E. All of the above   **(Schwartz-1646,1649)**

55. The main and the common symptom of vertebrobasilar insufficiency is:

A. Vertigo   B. Diplopia   **AMC 1986,1987**
C. Fainting   D. Nausea

56. Locked-in-syndrome is usually seen in lesions of the following except:

AMU 1989

A. Thalamus   B. Pons
C. Tegementum   D. Medulla

57. Which of the following is not an early feature of extramedullary spinal tumour :

AIIMS 1986; AMC 1986

A. Bladder involvement
B. Touch fibres on the side of lesion
C. Proprioceptive fibres involved on the side of lesion
D. Muscle weakness below lesion   **(Bailey-524)**

58. In Infants, papilloedema may be seen in :   **PGI 1984**
A. Frontal lobe tumors   B. Temporal lobe tumours
C. Medulloblastoma   D. Sagittal sinus thrombosis

59. Brain space occupying lesion causes death by:

Delhi 1984 1986, UPSC 1982; PGI 1983, 1987

A. Acute Hypertension   B. Brain herniation
C. Cushing syndrome   D. Hypotension   **(Bailey-552)**

60. The commonest of triad of brain SOL are all except :

PGI 1984

A. Headache   B. Vomiting
C. Diplopia   D. Papilloedema   **(Bailey-558)**

61. Disease which is congenital is: **AMU 1985**
A. Glioblastoma   B. Medulloblastoma
C. Multiple sclerosis D. Neurofibroma

Ans.: 53. C   54. E   55. A   56. A   57. A   58. D   59. B   60. C   61. C

62. **Tibulant ataxia is typically seen in:**   **AIIMS 1985**
    A. Medulloblastoma   B. Grave's disease
    C. Multiple sclerosis   D. Aortic regurgitation
63. **In children below 10 year, comonest cause of Pineal calcification is:**
    A. Tumour   B. Cysticercosis   **AIIMS 1985**
    C. Toxoplasmosis   D. Normal
64. **Pineal calcification is usually ——shaped:**   **AIIMS 1985**
    A. Spherical   B. Oval
    C. Comma   D. Diamond
65. **Below the level of the lesion in spinal concussion (produced by long axis stretch on the cord accompanying flexion) there is (with one exception):**
    A. A flaccid paralysis   **AIIMS 1986**
    B. Retention of urine
    C. Loss of joint sense
    D. Loss of pain and temperature sensation
66. **In the immediate management of patient who may possibly have cervical spinal injury which of the following statement is incorrect?**
                                                              UPSC 1985
    A. Cervical injuries should be transported with the head lifted on to a pillow
    B. The patient should be left lying on the stretcher until examination is completed
    C. If a cord lesion is present the best imme-diate examination is testing the sensory level to pink prick on the trunk
    D. A lateral X-ray should be taken without disturbing the patient
                                                              (Bailey-517)
67. **Ideal treatment to relieve pain after nerve suture is:**   **Delhi 1992**
    A. Exercise   B. Analgesics
    C. Narcotic analgesics   D. Sedatives
    E. Steroids
68. **Which is not a sign of increased intracranial pressure:**
    A. Bitemporal hemianopia   B. Headache   **AMU 1986**
    C. Bradycardia   D. Convulsions   **(Bailey-554)**
69. **The subclavian steal syndrome (giddiness associated with upper limb exercise) can be caused by:**   **DNB 1990**
    A. Stenosis of the origin of the vertebral artey
    B. Stenosis of the origin of the internal caro- tid artery
    C. Stenosis of the orign of the subclavian artery

Ans.: 62. C   63. A   64. C   65. C   66. A   67. A   68. A   69. C

D. Stenosis of the subclavian artery where it crosses first rib
(Bailey-213)

70. **Characteristics of the congenital hydrocephalus include each of the following except:**

A. Convulsion
B. Sun set sign   **AP 1989**
C. Crack-pot sign
D. Transillumination
(Bailey-554)

71. **Witzelsucht syndrome (i.e "Pathological Joking") is seen in :**
Rohtak 1986

A. Frontal lobe tumours
B. Parietal lobe tumours
C. Temporal lobe tumours
D. IV ventricular tumour

72. **Empty sella syndrome is often characterized by:**   **AMU 1986**
A. Pituitary tumour
B. Cretinism
C. Acromegaly
D. None of the above

73. **Protrusion of the lips occurring due tapping the skin at angle of the mouth (Escherich's sign) is seen in:**
AIIMS 1986

A. Frontal lobe damage
B. Tuberous sclerosis
C. Tetany
D. Hyperparathyroidism

74. **The incidence of thymoma in patient with myasthenia gravis is:**
AIIMS 1986

A. 30-45%
B. 20-30%
C. 15-20%
D. 5-15%

75. **Which of the following lesions of the thymus is most commonly associated with myasthenia gravis?**   **PGI 1987**
A. Lymphoepthelioma
B. Teratoma
C. Granulomatous thymoma
D. None of the above

76. **After an open injury,the optimum time for nerve suture is:**
PGI 1985

A. Immediately
B. Within one month
C. 1-2 months
D. 2-4 months
E. When wound is free from infection   (Bailey-378, 379)

77. **The following are true about the characterisitics of sacrococcygeal teratoma except :**   **CMC 1986**
A. One of the most common large tumours seen during first three months of life
B. Males are more often affected than females
C. Tumours arise between sacrum and rectum
D. Prone to become malignant   (Bailey-866)

Ans.: 70. A   71. A   72. D   73. C   74. A   75. A   76. E   77. B

78. **All of the following are indications for ventilatory support in patients with head injury except :**   **CSE 1995**
   A. PaO_2 less than 70 mm Hg
   B. $PaCO_2$ more than 45 mm Hg
   C. PaO_2 saturation less than 90%
   D. $PaCO_2$ less than 35 mm Hg

79. **Absolute 3rd Nv palsy occurs in :**   **TN 1998**
   A. Aneurysm of middle cerebral A
   B. Aneurysm of anterior cerebral A
   C. Aneurysm of posterior communicating A
   D. Aneurysm of inferior cerebellar A

80. **Brain abscess may be due to following :**   **AMC 1985**
   A. Chronic S.O.M.   B. Chronic lung abscess
   C. Trauma   D. Any of the above   **(Bailey-557)**

81. **The treatment of cerebral oedema includes :**   **AMC 1985**
   A. Restriction of fluid intak   B. Diuretics
   C. Glucocorticoids   D. Maintenance of airway
   E. All of the above

82. **Non metastatic neurological manifestation in bronchogenic carcinoma includes all the following, except :**   **CMC 1987**
   A. Cerebellar degeneration   B. Hoarseness of voice
   C. Myopathy   D. Peripheral neuropathy

83. **The neurosurgical procedure used in Parkinson's disease, electrocoagulation of a localized area with in the brain presumably blocks the output,Which in the form of an excessive feedback, originates in the:**   **UPSC 1085**
   A. Basal ganglia
   B. Motor cortex
   C. Precentral cortex
   D. Medullary portion of the reticular formation

84. **Treatment of choice for subgaleal hematoma:**   **UPSC 1987**
   A. Incision and evacuaton   B. Needle aspiration
   C. Antibiotics and then drai   D. Conservative

85. **Facial nerve palsy is seen in the following fracture:**   **PGI 1984**
   A. Anterior cranial fossa   B. Posterior cranial fossa
   C. Middle cranial fossa   D. Cranial vault

86. **Consider the following sites of obstruction :**   **CSE 1996**
   1. Foramina of Munro
   2. Outside the ventricular system
   3. At the exist of the fourth system

Ans.: 78. D   **79.** C   **80.** D   **81.** E   **82.** B   **83.** A   **84.** D   **85.** C   **86.** A

Hydrocephalus is described as non-communicating when the obstruction is at : CSE 1996

- A. 1,2 and 3
- B. 1 and 2
- C. 2 and 3
- D. 1 and 3

87. Subdural haemorrhage is commonly because of rupture of:
    - A. Middle cerebral artery   UPSC 1983, 1984, 1986;
    - B. Anterior cerebral artery   AIIMS 1984, 85, 87;
    - C. Posterior cerebral artery   UPSC 1984, 86; AI 1989
    - D. Dural sinuses   (Bailey-550)

88. The cause of slowly progressive paraplegia, root pain and patchy sensory loss is: AIIMS 1984. 1985
    - A. Motor neurone disease
    - B. Fluorosis
    - C. Leprosy
    - D. Cysticercosis

89. Which of the folowing is not a feature of midbrain lesion:
    AIIMS 1984, 85,88
    - A. Upper gaze palsy
    - B. Pupillary dilatation
    - C. Crossed hemiplegia
    - D. Extension of arm

90. Head injury with fracture of base of skull may develop:
    AIIMS 1984 85
    - A. Kernig's sign
    - B. Homan's sign
    - C. Papilloedema
    - D. CSF Rhinorrhoea

    (Bailey-549)

91. Most useful investigation in head injury is: AIIMS 1984, 85
    - A. X-ray lateral view
    - B. Angiogram
    - C. Ventriculography
    - D. CT scan   (Bailey-552)

92. A patient 45 year old had suboccipital headache, vomiting, confusion, death in 2 hours. Most common cause is: AIIMS 1985 86
    - A. Glioblastoma degeneration
    - B. Berry aneurysm rupture
    - C. Meningitis
    - D. Hydrocephalus   (Bailey-563)

93. Following are used to decrease intra cranial tension except :
    - A. Lasix
    - B. Glycerol   Rajasthan 1994
    - C. IV barbiturates
    - D. $PCO_2 > 45$ mmHg

    (Bailey-548)

94. The commonest supratentorial tumour in adults is:
    AIIMS 1985 ,86, 87 Delhi 1985, 86; UPSC 1984 AMC 1985
    - A. Meningioma
    - B. Glioma
    - C. Medulloblastoma
    - D. Craniopharyngioma

    (Bailey-549)

Ans.: 87. D   88. B   89. D   90. D   91. D   92. B   93. C   94. B

95. **Which of the following does not drain into cavernous sinus:**
   A. Superficial middle cerebral vein   **AIIMS 1985, 86**
   B. Ophthalamic vein
   C. Sphenoparietal sinus
   D. Great vein of Galen:

96. **A patient with 3rd nerve palsy had no pupillary abnormality. He has:**   **AIIMS 1988**
   A. Trauma
   B. Diabetes millitus
   C. Anterior cerebral artery aneurysm
   D. Pituitary tumour

97. **Features of posterior inferior cerebellar artery thrombosis includes all of the following except :**   **AMU 1990**
   A. Sudden onset of severe vertigo
   B. Acute cerebellar S/S with nystagmus to the side of lesion
   C. The involvement of 10,11,12, nerves
   D. Horner's syndrome

98. **The most commonly affected lower limbs muscle in poliomyelitis is:**   **AIIMS 1986,87**
   A. Quardriceps femoris   B. Tibialis anterior
   C. Tibialis posterior   D. Peroneus longus

99. **Following may be useful management of CSF rhinorrhoea except :**
   A. Nasal packing   **AIIMS 1986, 88**
   B. Repeated lumbar puncture
   C. Antibiotics
   D. Craniotomy and dural repair   **(Bailey-552)**

100. **Acute extradural hematoma may have following features except :**   **AIIMS 1986**
   A. Hematoma under temporalis muscle
   B. Presence of lucid interval
   C. Tear of superior cerebral vein
   D. Fracture squamous temporal bone   **(Bailey-550)**

101. **All are seen in ulnar nerve palsy except :**   **AIIMS 1986**
   A. Loss of sensation over medial aspect of palm
   B. Loss of sensation over anatomical snuff box
   C. Paralysis of interossei
   D. Postive froment sign

Ans.: 95. D   96. B   97. C   98. A   99. A   100. C   101. B

102. Artery commonly involved in cirsoid Aneurysm is :
  A. Superficial temporal artery   **UPSC 1985; Rajasthan 1994**
  B. Carotid artery
  C. Femoral artery
  D. Tibial artery

103. Best management of infected subaponeurotic hematoma is:
  A. Repeated needle aspiration   **AIIMS 1986**
  B. I and D (Incision and Drainage)
  C. Antibiotics alone
  D. Open repair

104. Hemiplegia is most often caused by thrombosis of :
                                              AIIMS 1986 1989

  A. Anterior cerebral artery   B. Posterior cerebral artery
  C. Middle cerebral artery   D. Basilar artery

105. Thalamectomy in parkinsonism relieves :   **NIMHANS 1997**
  A. Tremor   B. Rigidity
  C. Ataxia   D. Akinesia

106. Most common extra medullary toumour of spinal cord is:
                                        AMC 1987; AIIMS 1986, 87

  A. Neurofibroma   B. Meningioma
  C. Ependymoma   D. Metastatic tumour

107. In meningioma not seen is: **Delhi 1990**
  A. Lameller calcification
  B. ↑ Diploic space   **(Bailey-560)**
  C. Bone spicule in inner table
  D. Decalcification in inner table

108. Regarding cephalohaematoma: **AIIMS 1990**
  A. Is due to periosteal injury
  B. Focal swelling under periosteum
  C. Always associated with jaundice
  D. Many lead to cerebral palsy

109. Central area of face is called Dangerous area because :
                                                     AIIMS 1990

  A. Infection in this area causes cavernous sinus thrombosis
  B. This area is liable to tumour
  C. Fatal haemorrhage may occur
  D. Connected directly with middle ear

Ans.: 102. A   103. B   104. C   105. A   106. A   107. D   108. B
       109. A

110. Footdrop results because of injury to:

                                    PGI 1986; Delhi 1989; UPSC 1983
A. Superficial peroneal nerve   B. Deep peroneal nerve
C. Posterior tibial nerve   D. Anterior tibial nerve

111. Foot drop occurs due to lesion of all except :   PGI 1987
A. Sciatic nerve   B. Common peroneal nerve
C. L_5 root   D. S_1 root

112. Operative indications for paraplegia is: PGI 1987
A. Progressive motor loss inspite of conservative management
B. Loss of consciousness
C. No improvement in sensory loss within 1 week
D. Incontinence of urine

113. Which of the following is a cystic cranial tumour :   Kerala 1998
A. Craniopharyngioma   B. Astrocytoma
C. Medulloblastoma   D. Ependynoma

114. Bruising over mastiod process appearing a day or two after head injury indicating fracture of middle cranial fossa is: DNB 1993
A. Guerin's sign   B. Coleman's sign
C. Battle's sign   D. Babinski's sign

115. A patient presepts with hirning type of severe headache O/E scalp tenderness is severe. Most probabale diagnosis is :

                                    JIPMER 1997
A. Migraine   B. Temporal arteritis
C. Tension headache   D. Hypertension

116. Bilateral pyramidal signs develop early in:

                                    AMC 1986; PGI 1989
A. Cerebral hemorrhage   B. Subarachnoid hemorrhage
C. Pontine hemorrhage   D. Hemotomyelis

117. True about meningioma is all except :   PGI 1989
A. 19% of brain tumour
B. Parasagittal meningioma common
C. Reactive hyperostosis
D. Flat
E. Arises from meninges   (Bailey-560)

118. An elderly patient presents in the casualty with a head injury. Immediate step to be taken is :   Bihar 1998
A. Send for scanning   B. Establish IV access
C. Endotracheal intubation   D. Send blood for cross matching
                                                    (Bailey-552)

Ans.: 111. B   112. A   113. A   114. C   115. B   116. D   117. D
118. C

119. **Raised intracranial pressure is suspected if there is:**

    UPSC 1986;PGI 1990

    A. Unitateral headache   B. Ptosis

    C. Neck rigidity   D. Dilated pupil   **(Bailey-547)**

120. **Transient syndrome of vertigo, diplopiaslurred speech and paresthesia is:**

    AMC 1984, 1986

    A. Anterior communicating artery aneurysm

    B. Basilar artery insufficiency

    C. Middle cerebral artery thrombosis

    D. Posterior communicating artery aneurysm

121. **A 22-year old male is admitted with fracture of the left femur. Two days later, he becomes mildly confused, has a respiratory rate of 40/ min and scattered petechial rash on his upper torso. Chest X-ray shows patchy alveolar opacities bilaterally. His arterial blood gas analysis is abnormal. The most likely diagnosis is :**

    CSE 1998

    A. Cerebral oedema with early neurogenic pulmonary oedema

    B. Pulmonary thrombo-embolism

    C. Chest contusion

    D. Fat embolism

122. **Most common type of spina bifida is:**   **JIPMER 1984, 87**

    A. Meningocele   B. Meningo myelocele

    C. Myelocele   D. Spinal bifida occulta

    (Bailey-457)

123. **A scooter is hit from behind. The rider is thrown off and he lands with his head hitting the kerb. He does not move, complains of severe pain in the neck and in unable to turn his head. Well-meaning onlookers rush up to him and try to make him sit up. What would be the best course of action in this situation?**

    UPSC 1997

    A. He should be propped up and given some water to drink

    B. He should not be propped up but turned on his face rushed to the hospital

    C. He should be turned on his back and a support should be placed behind his neck and transported to the nearest hospital

    D. He should not be moved at all but carried to the nearest hospital in same position in which he hasbeen since his fall

---

Ans.: 119. D   120. B   121. D   122. D   123. D

124. **After complete division of a nerve, retrograde degeneration occurs as high as——node of Ranvier:**   **DNB 1991**
    A. 1st   B. 2nd
    C. 3rd   D. 4th
    E. 5th

125. **Meralgia parasethetica is an entrapment neuropathy of the:**
                                    Delhi 1986; AIIMS 1986
    A. Musculocutaneous nerve   B. Ilio inguinal nerve
    C. External cutaneous nerve   D. Lateral popliteal nerve

126. **Consider the following procedures :**   **CSE 1998**
    1. Ventriculoperitoneal shunt
    2. Ventriculocisternal shunt
    3. Ventriculopleural shunt
    Surgical procedures prescribed for the treatment of hydrocephalus would include :
    A. 1 and 3   B. 1 and 2
    C. 2 and 3   D. 1,2, and 3

127. **A child presnets with ataxia and incoordination. Nystagmus is observed on lateral gaze towards the right side. Areflexia and hypotonia are also present. The head is titled towards right side and the child walks with a broad base. The site of the lesion is :**
                                    UPSC 1997
    A. Cerebellum on the right side
    B. Cerebellum on the left side
    C. Brain stem on the left side
    D. None of the above

128. **Malignant astrocytoma is most common in:**   **DNB 1989**
    A. Frontal lobe   B. Temporal lobe
    C. Parietal lobe   D. Cerebellum

129. **Patient presents with high fever, signs of raised ICT and a past history of chronic otitis media. Likely diagnosis :**   **AIIMS 1981, 84**
    A. Brain abscess
    B. Pyogenic meningitis
    C. Acute subarachnoid haemorrhage
    D. Acute osteomyelitis of skull bone   **(Bailey-556)**

130. **Dumbell tumour is seen in:**   **AMC 1985**
    A. Meningioma   B. Neurofibroma
    C. Epidendymoma   D. Thymoma

Ans.: 124. A   125. C   126. B   127. B   128. A   129. B   130. B

131. **A-10 year old child presents with midline cerebellar tumour. Most likely diagnosis is :** AI 1994
    A. Medulloblastoma
    B. Astrocytoma
    C. Glioblastoma
    D. Hemangioblastoma

132. **The most malignant brain tumour is:** Karnataka 1987
    A. Glioblastoma multiforme
    B. Spongioblastoma
    C. Ependymoma
    D. Oligodendroglioma

133. **All are true about Glasgow coma scale except :** AIIMS 1994
    A. Consists of eye opening, motor and verbal response
    B. Score between 3-15 (Bailey-552)
    C. ↑ Score indicates poor prognosis
    D. Obeying motor command is given maxi-mum score

134. **Cerebral embolism is common in —— position:** AIIMS 1994
    A. Supine
    B. Tredelenburg's
    C. Prone
    D. Lateral

135. **Arnold Chiari malformation is:** UPSC 1994
    (Bailey-554)
    A. Agenesis of cerebellum
    B. Herniation of hind brain and cerebelum into cervical canal
    C. Arteriovenous malformation of cerebellum
    D. Healing usually occurs without skin grafting

136. **In sacral meningomyelocele, false is:** AI 1995
    A. Spasticity is feature of lower limbs (Schwartz-1924)
    B. Hydrocephalus is often seen
    C. Bladder incontinence is present
    D. Lax anal sphincter is present

137. **A brain tumour which has CSF metastasis and is radiosensitive:** AI 1995, 96
    A. Ependymoma
    B. Medulloblastoma
    C. Pinealoblastoma
    D. Astrocytoma
    (Schwartz-1888)

138. **Metastasis outside the brain occurs in brain tumour:**
    A. Crainopharyngioma
    B. Glioblastoma Kerala 1995
    C. Medulloblastoma
    D. Hemangioblastoma
    (Schwartz-1888)

139. **In person with rupture of intracranial aneurysm, deterioration in neurological condition after 24 hours may occour due to:** PGI 1994
    A. Rerupture
    B. Spasm
    C. Increased intracranial tension
    D. Hydrocephalus

Ans.: 131. A   132. A   133. C   134. A   135. B   136. D   137. B
138. C   139. B

140. Posteroinferior part of the internal capsule is supplied with :
                                                                  Jipmer 1997
   A. Charcot artery
   B. Huyhman's artery
   C. Anterior bowel of the posterior choroid artery
   D. Inferior choroidal artery

141. Neuroglia responsible for phagocytosis: **AIIMS 1994**
   A. Fibrous astrocytes   B. Protoplasmic astrocytes
   C. Oligodendrocytes   D. Microglia

142. Dysphasia is prominent in disorders of :   **AIIMS 1984**
   A. Dominant parietal lobe
   B. Non-parietal lobe
   C. Dominant temporal lobe
   D. Non-dominant temporal lobe

143. Which one of the folowing is diagnostic of complete transection of spinal cord after an injury of three weeks'duration: **UPSC 1995**
   A. The rectum of voluntary motor power below the level of the lesion
   B. The rectum of sensations below the levels of the lesion
   C. The rectum of reflex activity with full recovery of sensations below the lesion
   D. The rectum of reflex activity with partial recovery of sensations below the lesion   **(Bailey-508 to 510)**

144. A one month old female child has a swelling over the back in the sacral region .There is no cough impulse or erosion of the coccyx. The most likely clinical diagnosis would be: **UPSC 1985**
   A. Meningocele   B. Lipoma
   C. Sacro-coccgeal teratoma D. Neurofibroma

                                                          (Bailey-1121)
145. The most common tumour of pineal gland is: **AP 1993**
   A. Lipoma   B. Astrocytoma
   C. Haemangioma   D. Germinoma

146. Plexiform neurofibromatosis most commonly affects:
                                                    Delhi 1993; MP 1993
   A. Facial nerve   B. Trigeminal nerve
   C. Peripheral nerve   D. Glassopharyngeal nerve

147. Suture separation with proptosis is seen in :   **PGI 1995**
   A. Wilm's tumor   B. Neuroblastoma
   C. Multiple myeloma   D. Phaeochromocytoma

Ans.: 140. D   141. D   142. C   143. C   144. C   145. D   146. B
    147. B

148. **Commonest Presentation of spinal cord tumor is :** PGI 1996
    A. Pain
    B. Bruit over vertebral column
    C. Gait deformity
    D. All of the above **(Bailey-524)**

149. **Most common endocrine tumor of pituitary is :** PGI 1996
    A. GH Tumor
    B. ACTH tumor
    C. Prolactinoma
    D. TSH secreting tumor
    (Bailey-561)

150. **A lady presents with galactorrhoea and visual defects, investigation of choice is :** AI 1997
    A. Prolactin
    B. GH
    C. FSH
    D. LH **(Bailey-561)**

151. **Aneurysm is commonly seen in following except :**
    AI 1997
    A. Ant. cerebral artery
    B. Basilar artery
    C. Vertebral artery
    D. Post. communication artery
    (Bailey-563)

152. **Among following, CNS tumor with best prognosis is :**
    AIIMS 1994; AI 1994, 97
    A. Cerebral astrocytoma
    B. Cerebellar astrocytoma
    C. Medulloblastoma
    D. Glioblastoma **(Bailey-403)**

153. **Froment test is used for —— nerve :** Delhi 1996
    A. Ulnar
    B. Radial
    C. Median
    D. Axillary **(Bailey-539)**

154. **Aneurysm-8cm, commonest complication is :** Delhi 1996
    A. Thromboembolism
    B. Seizures
    C. Hypertension
    D. Hemiparesis

155. **Pituitary tumor produces :** Delhi 1996
    A. Bitemporal hemianopia
    B. Binasal hemianopia
    C. Unilateral quadrantopia
    D. Increase in blind spot only
    (Bailey-561)

156. **The pterion corresponds to the following except :**
    Karnataka 1996
    A. Anterior pole of insula
    B. Middle cerebral artery
    C. Transverse sinus
    D. Lateral cerebral sulcus

157. **Level of lesion in akinetic mutism is in :** Delhi 1996
    A. Pons
    B. Midbrain
    C. Cerebellum
    D. III ventricle

Ans.: 148. C   149. C   150. A   151. C   152. C   153. A   154. A
     155. C   156. C   157. D

158. **Nerve most commonly affected in intracranial subclinoid anterior aneurysm is :**   **AI 1996**
    A. II   B. III
    C. VI   D. VII   **(Bailey-563)**

159. **Non neoplastic compressive lesions of spinal cord are all except :**
    A. Inter vertebral disc prolapse   **Kerala 1996**
    B. Aneurysm of special vessels
    C. A-V malformation of the spinal vessels
    D. Arachnoiditis
    E. None of the above

160. **Cranial accessory nerve injury causes paralysis of :**   **AI 1996**
    A. Sternocleidomastoid   B. Stylopharyngeus
    C. Pharyngeal muscles   D. Levator scapulae

161. **Craniotomy is indicated in all except :**   **AI 1996**
    A. Depressed fracture   B. Recurrent CSF rhinorrhoea
    C. Growing fracture   D. Compound fracture

162. **Spinal cord compression is due to all except :**   **AI 1996**
    A. Lymphoma   B. Neurofibroma
    C. Ependymoma   D. Glioma

163. **Which of the following is not a cause of angiogenic cerebral oedema:**
    A. Normal pressure hydrocephalus   **Kerala 1998**
    B. Tumors
    C. Meningitis
    D. None

164. **The most common physical sign of cerebral metastasis is :**
                                                   Karnataka 1998
    A. Epilepsy   B. Focal neurological deficit
    C. Papillodema   D. Visual defects

165. **Blood brain barrier is not present in following except :**
                                                   Rajasthan 1998
    A. Neurohypophysis   B. Area postrema
    C. Subfornical organ   D. IV ventricle

166. **Match List I (Glasgow coma scale) with List II (Systolic blood pressure) and select the correct answer using the codes given below the Lists :**

List I	List II	**CSE 1998**
A. 13 to 15	1. 76 to 89	
B. 9 to 12	2. 50 to 75	
C. 6 to 8	3. 1 to 49	
D. 4 to 5	4. > 89	

Ans.: 158. B   159. D   160. C   161. A   162. A   163. A   164. B
          165. D   166. A

Codes :

	A	B	C	D
A.	4	1	2	3
B.	1	2	3	4
C.	4	3	2	1
D.	3	2	1	4

167. **Following an incised wound in the front of the wrist, the patient is unable to oppose the tips of the little finger and the thumb. The nerve involved is :** Orissa 1998
    A. Median nerve   B. Ulnar nerve
    C. Median and ulnar nerve   D. Radial and ulnar nerve

168. **Revised trauma score includes all of the following except :**
    A. Respiratory rate   B. Urinary output   **CSE 1999**
    C. Systolic blood pressure   D. Glasgow coma scale
    (Bailey-279)

169. **The correct order of priorities in the initial management of head injury is:** CSE 1999
    A. Airway, Breathing, Circulation, Treatment of extracranial injuries
    B. Treatment of extracranial injuries, Airway, Breathing, Circulation
    C. Circulation, Airway, Breathing, Treatment of extracranial injuries
    D. Airway, Circulation, Breathing, Treatment of extracranial injuries
    (Bailey-552)

170. **All of the following are the features of congenital hydrocephalus except :** CSE 1999
    A. Enlarged tense fontanelles
    B. Downward displacement of eyeballs
    C. Papilloedema
    D. Squint and nystagmus   (Bailey-553)

171. **A non-contrast CT scan in a person brought with hostory of head trauma showed a bi-convex lens shaped lesion. What is the diagnosis:** AIIMS 1999
    A. Sub-archnoid haematoma   B. Sub-dural haematoma
    C. Extradural haematoma   D. Brain edema   (Bailey-551)

172. **MC causative of infection in spinal epidural space is :**
    A. Proteus   B. Pseudomonas   **Kerala 1999**
    C. Klebsiella   D. Staph aureus

173. **In patients of head injuries with rapidly increasing intracranial tension without haematoma, the drug of choice for initial management would be :** UPSC 2000

Ans.: 167. A   168. B   169. A   170. C   171. C   172. D   173. C

A. Lasix   B. Steroids
C. 20% Mannitol   D. Glycine   **(Bailey-548)**

174. **Patient having hyperprolactinaemia (20 ng/ml serum) with inappropriate lactation should be investigated by: TNPSC 1998**
A. X-ray of sella turcica
B. CAT scanning of brain
C. Ophthalmoscopic examination   **(Bailey-561)**
D. All of the above

175. **A comatose patient with head injury at a P.H.C. before transferring to a higher centre, should have the following attended to :**
                                                       TNPSC 2000
I. Clearing mouth & throat and ensuring that tongue does not fall back
II. Catheterising the urinary bladder aseptically
III. Instituting anti-shock measures like I.V. fluids etc.
IV. Sedation using narcotics
Of these
A. All are correct   B. I and III correct
C. III only is correct   D. I, II and III are correct

176. **True about Berry-aneurysm is following except :   PGI 2000**
A. Associated with familial syndrome
B. Most common site of rupture apex which causes SAH
C. Wall contains S. muscle fibroblasts
D. 90% occur in ant. part of circulation at branching poins

177. **A patient has an accident with resultant transection of the pituitary what will not occur :   AI 2001**
A. Diabetes mellitus   B. Diabetes insipidus
C. Hyperprolactinemia   D. Hypothyroidism

178. **Patient of head injury, has no relatives, requires urgent cranial decompression; Doctor should :   AI 2001**
A. Operate without formal consent
B. Take police consent
C. Wait for relatives
D. Should not perform surgery

179. **Consider the following statements :   UPSC 2001**
Cerebral protection in head injury is provided through
1. Hypothermia at 35°C.
2. Hyperventilation
3. Administration of barbiturates
4. Administration of phencyclidine

Ans.: 174. D   175. B   176. C   177. A   178. A   179. B

Which of the above statements are correct ?
A. 1,2, 3 and 4   B. 1, 2 and 3
C. 2, 3 and 4   D. 1 and 4

180. Which one of the following cranial nerves is most often injured in patients with fracture of the middle cranial fossa ?   UPSC 2001
A. Sixth cranial nerve   B. Eighth cranial nerve
C. Tenth cranial nerve   D. Eleventh cranial nerve

181. A patient is brought with head injury, head on collision and BJP 90/60. Tachycardia present; diagnosis is :   AI 2001
A. Extradural hematoma   B. Subdural hematoma   (A-21)
C. Intracranial hemorrhage   D. Intraabdominal bleed

182. Characteristic finding in CT Scan of a TB case :   AI 2001
A. Exudates seen in basal cistern
B. Hydrocephalus is commonly seen
C. Tuberculomas are often calcified
D. CT is diagnostic of TBM

183. Stereolactic radiosurgery is done for :   JIPMER 2002
A. Glioblastoma multiforme   B. Medullo blastoma spinal cord
C. Opendyomoma   D. AV malformation of brain
                                                    (CSDT-858)

184. Which of the following requires emergency operation in setting without tertiary care facilities :   SGPGI 2002
A. Extradural hemorrhage   B. Subdural hemorrhage
C. Subarachnoid hemorrhage   D. Intacerebral hemorrhage
                                                    (Bailey-550)

185. Which of the following tumors is common in extramedullary intra-dural location :   SGPGI 2002
A. Ependymoma   B. Metastasis
C. Astrocyotoma   D. Neurofibroma   (CSDT-829)

186. Post traumatic increase in ICT, the following drug is not used :   JIPMER 2002
A. Mannitol   B. Frusemide
C. Dexamethasone   D. Glycerol   (CSDT-821)

187. All the following are correct about radiologic evaluation of a patient with Cushing's syndrome except :   AI 2002
A. Adrenal CT scan distinguishes adrenal cortical hyperplasia from an adrenal tumor
B. CT of sella tursica is diagnostic when a pituitary tumor is present

Ans.: 180. B   181. D   182. A   183. A   184. A   185. D   186. C
      187. B

   C. MRI of the adrenals may distinguish adrenal adenoma from carcinoma

   D. Petrosal sinus sampling is the best way to distinguish tumor from an ectopic ACTH producing tumor

(Bailey-561)

188. **A 24 year old man falls to the found when he is struck in the right temple by a base ball. Shile being driven to the hospital he lapses into coma. He is unresponsive with a dialated right pupil when he reaches the emergency department. The most appropriate step in initial cmanagement is :**

   A. CT scan of the head   **AI 2002**

   B. Craniotomy

   C. Doppler ultrasound examination of the neck

   D. X-rays of the skull and cervical spine.   **(Bailey-551)**

189. **The following are the clinical features of raised intracranial tension except :**   **UPSC 2002**

   A. Headache   B. Insomnia

   C. Bradycardia   D. Papilloedema

190. **A 10 year old child presented with headache, vomiting, gait instability and diplopia. On examination he had papilloedema and gait ataxia. The most probabale diagnosis is :**   **AIIMS 2002**

   A. Hydrocephalus   B. Brain stem tumour

   C. Suprasellar tumour   D. Midline posterior fossa tumour

191. **Which of the following would distinguish hydrocephalus due to aqueductal stenosis when compared to that due to Dandy-Walker malformation :**   **AIIMS 2002**

   A. Third ventricle size   B. Posterior fossa volume

   C. Lateral ventricular size   D. Head circumference

192. **The lumbar puncture was done in a patient with raised intracranial tension. The patient died suddenly on the table. The cause of death is most likely to be :**   **UPSC 2003**

   A. Middle cerebral artery hemorrhage

   B. Tentorial herniation

   C. Rupture of an aneurysm

   D. Loss of CSF

193. **Patient sustains head-injury; and develops quadriplegia and respiratory distress. He is suffering from :**   **UPSC 2003**

   A. Cervical vertebral fracture dislocation

   B. Brainstem hemorrhage

Ans.: 188. B   189. B   190. D   191. B   192. B   193. B

C. Cerebral laceration
D. Cerebral contusion

194. **Refsum's disease is a form of :**   **AIIMS 1986**
   A. Autonomic neuropathy   **(Harrison-2513)**
   B. Retinal degeneration
   C. Lepromatous leprosy
   D. Inborn error of metabolism

195. **Bell's palsy is due to :**   **Delhi 1983**
   A. A lesion at the geniculate ganglion   **(Harrison-2412)**
   B. Pressure and oedema in the facial canal
   C. Cerebellopontine angle tumour
   D. Medullary infarction

196. **Horner's syndrome components include following except :**
                                                    Delhi 1985, 90

   A. Enophthalmos   B. Ptosis and Miosis
   C. Exophthalmos   D. Impaired sweating
                                         (Harrison-165, 175)

197. **The commonest form of intracranial neoplasm is :**   **Delhi 1989**
   A. Meningioma   B. Acoustic neuroma
   C. Glioma   D. Angioma   **(Harrison-2442)**

198. **Albuminocytological dissociation in CSF is observed in :**
                                              UPSC 1984, 86; AI 1997

   A. Viral meningitis   B. Cervical spondylosis
   C. Cerebral malaria   D. Infective polyneuritis
                                         (Harrison-2415)

199. **Primary amoebic meningo-encephalitis is a feature :**
                                                    Rohtak 1988

   A. Of extension of entamoeba from liver or intestine
   B. Largely treatable with antiamoebic drugs
   C. Due to free living amoeba flagellates
   D. Largely reversible   **(Harrison-1202)**

200. **In a patient with RHD angiography showed complete occlusion of a
   small branch of the middle cerebral artery with no other significant
   findings. Which of the following is the most likely cause of the
   woman's cerebrovascular accident ?**   **Manipal 1993**
   A. Arteriovenous malformation
   B. Atrial fibrillation with mural thrombus
   C. Berry aneurysm
   D. Deep vein thrombosis

Ans.: 194. D   195. B   196. C   197. C   198. D   199. C   200. B

201. **All of the following statements regarding the oculomotor nerve are true except :**   **UPSC 1998**
    A. It accommodates the eye   B. It raises the upper eyelid
    C. It innervates lateral rectus   D. It constricts the pupil

202. **In dystrophia myotonica, untrue is :**   **Delhi 1984**
    A. Autosomal dominant
    B. Cataracts
    C. Mental retardation
    D. Always presents with myotonia
    E. Glucose intolerance   **(Harrison-2351)**

203. **Hemianopia, cortical blindness, amnesia and thalamic pain are associated with the occlusion of :**   **UPSC 1998**
    A. Anterior cerebral artery
    B. Middle cerebral artery
    C. Posterior cerebral artery
    D. Basilar artery   **(Harrison-2363)**

204. **The commonest neurological disorder in elderly people :**
                                                        Nimhans 1987
    A. Trauma to the head   B. Brain tumour
    C. Stroke   D. Cerebral abscess
                                                        (Harrison-2396)

205. **Hypophysectomy is followed by all of the following consequences except:**
    A. Loss of axillary and pubic hair   **Manipal 1995**
    B. Loss of libido
    C. A greater degree of hypoglycaemia following an insulin injection
    D. A decrease in the plasma concentration of aldosterone

206. **Myoclonus is characterized by :**   **TN 1990**
    A. Shock like contractions
    B. Slow writhing irregular movements
    C. Quasi purposive non repetitive movements
    D. Alternating movements   **(Harrison-126)**

207. **An 80-year old person has started forgetting the names of familiar persons and places. There has been no confabulation. He tends to forget whether he has had his meals. Clinical and neurological examination reveal no abnormality. CT scan of the brain showed symmetrical enlargement of lateral ventricles and wider sulci. The most likely diagnosis is :**   **UPSC 1998**

Ans.: 201. C   202. D   203. C   204. C   205. D   206. A   207. B

A. Confusional state
B. Alzheimer's disease
C. Alcohol dementia
D. Chronic cerebrovasicular insufficiency   **(Harrison-2393)**

208. **The combination of polyneuritis, confusion, disorientation, loss of memory and tendency to confabulate is most likely due to :**
                                                              Bihar 1991

    A. Alcoholism   B. Pernicious anaemia
    C. Dermatomyositis   D. Charcot-Marie-Tooth disease
                                                              (Harrison-2416)

209. **Periodic paralysis may be associated with all of the following except:**
                                                              UPSC 1998

    A. Hyperkalaemia   B. Hypokalaemia
    C. Hyperthyroidism   D. Hypothyroidism
                                                              (Harrison-2538)

210. **Neurocysticercosis is diagnosed by :**   **UPSC 1987**
    A. Pneumoencephalography   B. Angiography
    C. EEG   D. MRI Scan .   **(Harrison-1248)**

211. **Motor neurone disease is thought to be related to :**
                                                              AIIMS 1984

    A. Slow virus infection   B. Porphyria
    C. Tetanus   D. Chromosomal aberration
                                                              (Harrison-2412)

212. **Anton syndrome is :**   **UP 1998**
    A. Cortical blindness   B. Patient denies that he is blind
    C. Both of the above   D. None

213. **Painful ophthalmoplegia is :**   **AIIMS 1984**
    A. Largerly reversible   B. A manifestation of migraine
    C. Related to intracranial tumor D. Irreversible   **(Harrison-177)**

214. **Huntington's sign is present in :**   **DNB 1995**
    A. Huntington's disease   B. Upper motor neurone disease
    C. Congenital syphilis   D. Parkinsonism

215. **Which of the following is X-linked recessive :**
                                                   UPSC 1985, 89; Delhi 1993

    A. Friederich's ataxia
    B. Huntington's chorea
    C. Peroneal muscular dystropathy
    D. Muscular dystropathy (Duchenne)   **(D-18)**

Ans.: 208. A   209. D   210. D   211. A   212. C   213. A   214. B
         215. D

216. Normal CSF pressure in sitting posture is ——— mm Hg :

Delhi 1993

   A. 4-6   B. 12-14

   C. 18-20   D. 20-23   **(Harrison-2492)**

217. Most common CNS manifestation of HIV infection is :

Jipmer 1993

   A. Acute meningitis   B. Encephalopathy

   C. Vacuolar myelopathy   D. Dementia   **(Harrison-151)**

218. Right superior quadrantanopia is caused by a lesion of :

   A. Right temporal lobe   B. Left temporal lobe   **AP 1990**

   C. Right parietal lobe   D. Left parietal lobe

219. Balaclava helmet type of sensory loss over the face is characteristic of :

TN 1990

   A. Syringomyelia   B Multiple sclerosis

   C. Tabes dorsalis   D. All of the above

220. L.P. is dangerous in :

AMC 1985

   A. Sub-arachnoid haemorrhage B. Cerebral tumor

   C. Cerebral thrombosis   D. Hypertensive encephalopathy

   `E. Encephalitis   (Harrison-2331)**

221. Choreiform movements, dementia in adult life and similar symptoms in family members makes the diagnosis of :   **Delhi 1983, 89**

   A. Acute chorea   **(Harrison-2397)**

   B. Habit spasms

   C. Huntington's chorea

   D. Dystonia musculorum deformans

222. Commonest cause of Intracerebral Haemorrage in adults :

AI 1993

   A. Hypertension   B. Ruptured berry aneurysm

   C. CVA   D. Trauma   **(Harrison-2371)**

223. Commonest cause of meningitis of fungal origin :   **AP 1996**

   A. Cryptococcus   B. Histoplasma

   C. Candida   D. Aspergillus   **(Harrison-1175)**

224. The characteristic feature of Parkinson's syndrome consist of :

AIIMS 1985; AI 1997

   A. Tremor, rigidity, hyperkinesia

   B. Tremor, rigidity, hypokinesia

   C. Rigidity, hypokinesia, tremor

   D. Chorea, athetosis, hypokinesia

   E. None of the above   **(Harrison-2399)**

Ans.: 216. A   217. D   218. B   219. A   220. B   221. C   222. B

     223. A   224. B

225. **Cerebral malaria may present with except :** UPSC 1989
    A. Normal CSF picture   B. Cerebellar features
    C. Convulsions   D. None of the above
    (Harrison-1206)

226. **A patient with hypoxaemia, and polycythaemia is able to restore his blood gases to normal by voluntary hyperventilation. The primary pathology is likely to be located in :** UPSC 1986
    A. Cerebellum   B. Cerebral cortex
    C. Bone marrow   D. Respiratory centre in medulla

227. **True statement regarding neurological manifestation of AIDS is :**
    Manipal 1998
    A. Dementia and delirium are most common CNS manifestations
    B. Neurological lesions do not develop until 2-3 years after HIV infection
    C. 25% of AIDS patients show subnormal intelligence on clinical examination
    D. CT Scan is diagnostic in a majority   (Harrison-1891)

228. **Which of the following is not associated with peripheral neuropathy?**
    DNB 1990
    A. Urticaria pigmentosa   B. Diabetes mellitus
    C. Nitrofurantion   D. Infectious mononucleosis
    E. Diphenyl hydantion   (Harrison-2498)

229. **The following are characteristics of amyotrophic lateral sclerosis except:**
    Delhi 1987
    A. Sensory signs do not occur
    B. Bladder never involved
    C. Relentless progression inevitable
    D. Associates with external ophthalmoplegia   (Harrison-2412)

230. **Following are features of right hemiparesis except :**
    JIPMER 1998
    A. Left to right disorientation   B. Acalculia
    C. Neglect of paralysed side   D. Gaze palsy   (Harrison-121)

231. **Which of the following is not a common complication of phenytoin therapy ?** UPSC 1990
    A. Photophobia   B. Drowsiness
    C. Diplopia with nystagmus   D. Morbiliform eruption
    (Harrison-2364)

Ans.: 225. D   226. D   227. A   228. A   229. D   230. A   231. A

232. Which is not a cause of acute coma :   **JIPMER 1998**
   A. Carotid A. Ischemia   **(Harrison-134)**
   B. Sub-arachanoid haemorrhage
   C. Cerebellar haemorrhage
   D. Basilar artery thrombosis

233. Brown syndrome resembles :   **CMC 1998**
   A. III nerve palsy   B. IV nerve palsy
   C. VI nerve pals   D. Internuclear Ophthalmoplegia
   E. Inferior oblique palsy

234. A 75-year-old woman is admitted to the hospital after have suffered a cerebrovascular accident. Physical examination reveal responding to any visual/auditory/tactile stimuli on the left side of her body. She is also found to have a deficit invoving only the inferior portion of the left visual field in both her eyes. CT scan shows a non-haemorrhagic infarction in the right hemisphere. Which lobe of this patient's brain has been principally affected by the stroke?
    Manipal 1993
   A. Frontal   B. Occipital
   C. Parietal   D. Temporal

235. All the drugs mentioned below cause peripheral neuropathy except:
    DNB 1989
   A. I.N.H   B. Vincristine
   C. Methotrexate   D. Cloxacillin   **(Harrison-435)**
   E. Emetine

236. Neurotoxicity of I.N.H. is related to :   **Bihar 1989**
   A. The dose of the drug
   B. Slow inactivation of drug   **(Harrison-1018)**
   C. Rapid inactivation of the drug
   D. None of the above

237. The commonest site of Berry aneurysm formation is :
   A. Vertebrobasilar territory   **UPSC 1988**
   B. Origin of post. communication artery from internal carotid
   C. Ant. Communicating artery
   D. Any of the above

238. The most common symptom of vertebrobasilar insufficiency is :
    Kerala 1990
   A. Hemianopic visual loss   B. Hemiparesis
   C. Vertigo   D. Diplopia   **(Harrison-2373)**

Ans.: 232. A   233. B   234. C   235. D   236. B   237. C   238. C

239. **The acute development of chorea requires the consi-deration of which of the following disorders ?**   **PGI 1986**
    A. Rheumatic fever   B. Tumors   **(Harrison-1340)**
    C. Systemic lupus erythematosus D. CO poisoning
240. **All of the following are true of multiple sclerosis, except :**
    A. Diplopia is common   **Delhi 1983**
    B. The g-globulin fraction of spinal fluid protein is increased
    C. Headache and aphasia are common
    D. Ataxia is common   **(Harrison-2455)**
241. **Which of the following predispose cerebral haemorrhage ?**
    A. Leukaemia   **AMC 1985**
    B. Thrombocytopenia   **(Harrison-2385)**
    C. Hypertensive vascular disease
    D. Anticoagulants
    E. All of the above
242. **Myotonic muscular dystrophy is different from other dystrophies in that:**
                                    AMC 1984
    A. Distal muscle weakness occurs before proximal weakness
    B. Cranial nerve involvement occurs
    C. Myotonia is present
    D. Cataract, frontal baldness and testicular atophy occur
    E. All of the above   **(Harrison-2518)**
243. **The commonest site of lesion in hemiplegia is :**   **AMU 1987**
    A. Frontal lobe   B. Cerebellum
    C. Internal capsule   D. Spinal cord   **(Harrison-121)**
244. **In Parkinsonism the drug of choice is :**   **DNB 1989**
    A. Methyl dopa   B. Levo dopa
    C. Anticholinergic drugs   D. Adrenergic drugs
                                    (Harrison-2399)
245. **Muscular atrophy particulary of the small muscles of the hand, weakness, and fasciculations, in the absence of sensory disturbances suggest :**
                                    UPSC 1982, 86, 90
    A. Syringomyelia   **(Harrison-2412)**
    B. Haematomyelia
    C. Amyotrophic lateral sclerosis
    D. Multiple sclerosis
    E. Poliomyelitis

Ans.: 239. A   240. C   241. E   242. E   243. C   244. B   245. C

246. **A patient after vomiting several times develops carpopedal spasm. The most appropriate treatment would be :**   **UPSC 1998**
    A. Intravenous injection of 20 ml 10% calcium gluconate solution
    B. Intravenous infusion of isotonic saline
    C. Oral ammonium chloride 2 gm four times a day
    D. 5% CO_2 inhalation

247. **Following are causes of post lumbar puncture headache, except :**
                                                    AIIMS 1986
    A. Dehydration   **(Harrison-73, 2331)**
    B. Altered C.S.F. pressure
    C. Rapid change in blood volume
    D. Early ambulation

248. **Trigeminal neuralgia is sometimes associated with :**   **DNB 1990**
    A. Motor neurone disease   **(Harrison-2420)**
    B. Syringomyelia
    C. Multiple sclerosis
    D. Subacute combined degeneration of cord
    E. Transverse myelitis

249. **The most important features of cerebellar ataxia is :   UPSC 1982**
    A. Dysdiadokokinesia   B. Positive Romberg's sign
    C. Muscular hypotonia   D. Motor ataxia   **(Harrison-127)**

250. **Which of the following is/are positive in case of pyramidal lesion ?**
                                                    AIIMS 1985
    A. Babinski's sign   B. Oppenheim's sign
    C. Gordon's sign   D. Chaddock's sign
    E. All of the above

251. **Which of the following gaits is found characteristically in cerebellar coordination ?**   **AIIMS 1983**
    A. 'Spastic' gait   B. 'Stamping' gait
    C. 'Reeling gait   D. Wadding gait
    E. Prancing gait (high-stepping gait)   **(Harrison-127)**

252. **Albumino-cytologic dissociation occurs in the CSF in :   PGI 1984**
    A. Spinal cord tumour   B. Poliomyelitis
    C. Neurosyphilis   D. Spinal cord section
    E. None of the above   **(Harrison-2415)**

253. **Lasegue's sign is seen in :**   **PGI 1986**
    A. Nerve root pressure   B. Intracranial lesions
    C. Myasthenia gravis   D. Transection of spinal cord

Ans.: 246. B   247. D   248. C   249. A   250. E   251. C   252. A
       253. A

254. Bromocriptine is used to treat : AMC 1989
   A. Myxedema
   B. Parkinsonism (Harrison-2399)
   C. Medullary carcinoma thyroid
   D. Obesity

255. The commonest cause of stroke in elderly is :
PGI 1983; AMC 1985, 87
   A. Embolism
   B. Thrombosis
   C. Berry aneurysm rupture
   D. Atrial fibrillation
(Harrison-2369)

256. Transient syndrome of vertigo, diplopia, slurring speech and paresthesia is : PGI 1983; AMC 1987
   A. Basilar artery insufficiency
   B. Anterior communicating artery aneurysm
   C. Middle cerebral artery thrombosis
   D. None of the above (Harrison-2370)

257. In Brown Sequard syndrome, true is :
Delhi 1983, 84; AMC 1985, 86
   A. No sensory loss
   B. Ipsilateral corticospinal signs
   C. Flexor plantar reflexes
   D. Contralateral flaccid-paralysis (Harrison-121,2436)

258. In TB meningitis, characteristic CSF finding is :
AIIMS 1982, 83, 85, 88; AMC 1985, 86, 89
   A. Low CSF sugar
   B. Xanthochromia
   C. Polymorphonucleocytosis
   D. Persistent proteinemia
(Harrison-1028)

259. Site of action of blocking immunoglobulin in Myasthenia gravis is :
AI 1988
   A. Motor end plate
   B. Presynaptic vesicles
   C. Synapse
   D. Cell body (Harrison-2515)

260. Mononeuritis multiplex is a feature of : AI 1989
   A. Polyarteritis nodosa
   B. Hypersensitive vasculitis
   C. Leprosy
   D. All of the above
(Harrison-1932)

261. Which is a feature of classical migraine : AI 1989
   A. Symptoms are better with increasing age
   B. No aura

Ans.: 254. B   255. B   256. A   257. B   258. A   259. A   260. D
   261. A

    C. Early treatment absorbs an attacks

    D. Does not respond to Ergot   **(Harrison-2351)**

262. **Hemiballism is caused by lesions of the :**   **AIIMS 1980**

    A. Caudate nucleus

    B. Contralateral subthalamic nuclei

    C. Putaman

    D. Substantia nigra   **(Harrison-125)**

263. **The most likely diagnosis in a seven-year-old girl who has periods of inattention and poor performance is :**   **Manipal 1996**

    A. Complex partial seizures   B. Grand mal epilepsy

    C. Jacksonian epilepsy   D. Petit mal epilepsy

264. **The commonest cause of peripheral neuropathy is :**   **AIIMS 1982**

    A. Leprosy   B. Diabetes mellitus

    C. Drugs   D. Alcohol   **(Harrison-2498)**

265. **Commonest cause of systemic autonomic neuropathy in the world is:**   **Delhi 1982; AIIMS 1983**

    A. Leprosy   B. Diabetes mellitus

    C. Chagas disease   D. Drugs   **(Harrison-2506)**

266. **The commonest cause of spastic paraplegia is :**

                Delhi 1981; AIIMS 1982; PGI 1982

    A. TB   B. Spinal tumours

    C. Trauma   D. Fluorosis   **(Harrison-2440)**

267. **Neuropathy mainly involves distal group of muscles except :**

                AIIMS 1982; PGI 1983, 88

    A. Diabetes mellitus   B. Alcoholic neurpathy

    C. Guilain Barre syndrome   D. Leprosy   **(Harrison-2507)**

268. **Myopathy mainly involves proximal muscle groups except :**

                Delhi 1982, 83

    A. Myotonia congenitia   B. Myotonia atrophica

    C. Drug induced   D. All of the above

                (Harrison-2507)

269. **Steroid induced myopathy mainly involves :**   **PGI 1981, 83**

    A. Shoulder   B. Pelvis

    C. Neck   D. Arms   **(Harrison-2527)**

270. **Wernicke's encephalopathy is due to deficiency of Vitamin :**

              PGI 1982, 86; UPSC 1985; ESI 1988, 90

    A. B_1 B. B_2

    C. B_6 D. B_{12} **(Harrison-2418)**

Ans.: 262. B   263. D   264. A   265. B   266. A   267. C   268. B

       269. B   270. A

271. The common cause of unilateral central scotoma include all the following except : Manipal 1996
- A. Glaucoma
- B. Optic nerve glioma
- C. Papillitis
- D. Parasellar meningioma

272. About brain death, all are true except :

Delhi 1986, 89; DNB 2001
- A. Isoelectric EEG (Harrison-5)
- B. Heart and lungs may by working
- C. Patient may be revived
- D. Follows after heart or long functioning stops

273. All cause optic neuritis except : AIIMS 1988, AI 1993
- A. Systemic steroids
- B. Ethambutol (Harrison-432)
- C. Chloroquine
- D. Quinolones

274. A patient had sudden severe headache followed by coma and hemiplegia. The C.S.F. was haemorrhagic. The blood pressure was 120/80 mm Hg and the pulse rate was 100/minute. Which one of the following drugs is indicated for this patient ? CSE 1995
- A. Nimodipine
- B. Amlodipine
- C. Ticlopidine
- D. Nifedipine

275. Which one of the following treatments was not indicated in this case?

CSE 1995
- A. Rest
- B. Morphine
- C. Aspirin
- D. Digitalis

276. Elevation of CPK occurs in : Delhi 1985, 88
- A. Ulcerative colitis
- B. Typhoid
- C. Muscle dystrophy
- D. Polio (Harrison-1681)

277. Which one of the following signs is most valuable in the diagnosis of Wernicke's encephalopathy? UPSC 1998
- A. Ataxia (Harrison-2496)
- B. Confusional state
- C. Korsakoff's psychosis
- D. Bilateral symmetrical ophthalmoplegia

278. Cerebellar ataxia is a complication of : Karnataka 1998
- A. Chicken pox
- B. Measles (Harrison-127)
- C. Herpes simplex virus
- D. Varicella virus

279. The commonest site of cerebral hemorrhage in hypertension is :

PGI 1988
- A. Pons
- B. Cerebrum (Harrison-2370)
- C. Putamen
- D. Thalamus

Ans.: 271. D   272. C   273. A   274. A   275. D   276. A   277. D
278. A   279. C

280. Wadding gait is seen in :   **AMC 1988**
    A. Cerebellar lesions   B. Alcoholism   **(Harrison-129)**
    C. Thalamic lesions   D. Muscular dystrophy

281. Temperature regulation is mostly controlled by :   **AMC 1987**
    A. Pons   B. Medulla   **(Harrison-130)**
    C. Thalamus   D. Hypothalamus

282. Hutchinson's pupils is :   **Delhi 1984**
    A. Seen in syphilis   **(Harrison-177)**
    B. Unilateral constricted pupil
    C. Irregular pupil
    D. Argyl Robertson pupil

283. History of transient ischemic attack, excludes :   **Delhi 1984**
    A. Embolic episodes   B. Thrombotic episodes
    C. Haemorrhage   D. Completed stroke
                                      (Harrison-2370)

284. Fasciculations are seen in all except :   **Delhi 1985, 87**
    A. Pseudomuscular hypertrophy   **(Harrison-131)**
    B. Motor Neuron disease
    C. Polymyositis
    D. Poliomyelitis

285. Amantadine in Parkinsonism acts by :   **Delhi 1985, 86**
    A. ↑ Dopamine level   **(Harrison-2399)**
    B. ↓ Cholinergic level
    C. ↑ Norepinephrine from adrenal
    D. None of the above

286. In Kernicterus, not seen in :   **Delhi 1986**
    A. Hydrocephalus   B. Athetosis   **(Harrison-1717)**
    C. Brown staining of teeth   D. Mental retardation

287. Drug contraindicated in Myaesthenia gravis is :

                                           Delhi 1986
    A. Pyridostigmine   B. Neostigmine
    C. Quinine   D. Ephedrine   **(Harrison-2512)**

288. A patient was clinically diagnosed as having meningitis. CSF findings did not suggest any abnromality on biochemical examination. Subsequently, the patient died. Postmortem examination confirmed the diagnosis of meningitis. Which one of the following infections was responsible for the death?   **CSF 1995**
    A. Viral infection   B. Cryptococcal infection
    C. HIV   D. H. influenzae

Ans.: 280. D   281. D   282. B   283. C   284. A   285. A   286. A
      287. C   288. B

289. **Common causes of stupor and coma include :**   AMC 1986
    A. Infection   (Harrison-133)
    B. Metabolic lesions
    C. Intoxication with alcohol and drugs
    D. Supra and subtentorial lesions
    E. All of the above

290. **Pseudotumour cerebri is seen in with the following, except :**
                                                          UPSC 1986, 94
    A. Gentamicin   B. Tetracycline
    C. Oral contraceptives   D. Hypervitaminosis A
                                                          (Harrison-435)

291. **Degenerative changes are seen in Huntington's chorea in :**
                                                              AMC 1986
    A. Cerebellum   B. Caudate nucleus
    C. Red nucleus   D. Pons

292. **Pseudocoma results from infarction or haemorrhage in :**
                                                              CSE 1998
    A. Pons   B. Mid-brain
    C. Medulla   D. Hypothalamus

293. **An old patient of tubercular meningitis is not respondito treatment.
    A new combination of drugs has to be prescribed for a prolonged
    period. Which one of the following should not be prescribed in this
    case ?**
                                                              CSE 1998
    A. Rifampicin   B. Ofloxacin
    C. Dehydrostreptomycin   D. Pyrazinamide

294. **Migraine causes the following except :**   AMC 1984
    A. Paraesthesia   B. Blurring of vision
    C. Dysphagia   D. Seizures   (Harrison-2351)

295. **Thrombosis of Anterior cerebral Artery, distal to the communicating
    branch leads to :**   Delhi 1992
    A. Contralateral Hemiparesis   (Harrison-2372)
    B. Ipsilateral hemiparesis
    C. Incontinence
    D. Seizures

296. **Fast knee Jerks but absent ankle Jerks may be seen in :**
                                                              UPSC 1989
    A. Motor neuron disease   B. Tabes dorsalis
    C. Tabo paresis   D. Syringomyelia (Harrison-1047)

Ans.: 289. E   290. A   291. B   292. A   293. C   294. C   295. C
    296. B

297. In temporal lobe epilepsy, not seen is :   **UPSC 1989**
   A. Dreamy states   **(Harrison-2355)**
   B. Visual and olfactory hallucinations
   C. Familiarity
   D. Fortification spectra

298. A 25 year old while pushing a car developed sudden headache, neck pain and positive Kernig's sign. BP-150/100 mm Hg and in fundus, subhyaloid haemorrhage. Diagnosis is :   **PGI 1983; UPSC 1989**
   A. Cerebral haemorrhage   B. Cerebrovascular accident
   C. Subarachnoid haemorrhage   D. Meningitis   **(Harrison-2388)**

299. A young man after taking alcohol slept and on waking up, he is unable to hold cup. His BP is 150/100 mm Hg. Correct diagnosis is :   **UPSC 1986, 89**

   A. Alcoholic neuropathy   B. Radial nerve palsy
   C. Ulnar nerve palsy   D. Transient ischaemic attack
                                                           (Harrison-2369)

300. A patient with motor neuron disease suddenly developed loss of emotional control and inappropriate laughing and weeping. Correct diagnosis is :   **UPSC 1989**
   A. Agitated depression   B. Schizophrenia
   C. Pseudobulbar palsy   D. Subarachnoid hemorrhage
                                                           (Harrison-2411)

301. Meningitis is caused by following except :
                                          PGI 1984; UPSC 1989
   A. Paracoccidiomycosis   B. Coccidiomycosis
   C. Systemic candidiasis   D. Cryptococcosis
                                                           (Harrison-2477)

302. Ipsilateral 3rd nerve palsy with contralateral hemiplegia is :
                                                           PGI 1983
   A. Millard Gobbler syndrome   B. Weber's syndrome
   C. Foville syndrome   D. Benedicts syndrome
                                                           (Harrison-121)

303. Palmomental reflex is seen in lesions of :   **AIIMS 1998**
   A. Frontal lobe   B. Parietal lobe
   C. Temporal lobe   D. Occipital lobe

304. Headache in office could be due to :   **Rohtak 1996**
   A. Myopia   B. Decreased illumination
   C. Tension   D. Any of the above
                                                           (Harrison-71)

Ans.: 297. D   298. C   299. D   300. C   301. A   302. B   303. A
        304. D

305. The "Oculo-cerebro-renal syndrome of Lower" comprises of following except :   AMC 1982
A. Cataract
B. Mental deficiency
C. Tubular defects of kidneys
D. Deafness

306. Signs of Horner's syndrome may occur in each of the following conditions, except :   AIIMS 1985
A. Syringobulbia   (Harrison-2377)
B. Syringomyelia
C. Thrombosis of posterior inferior cerebellar artery
D. Bronchial carcinoma at apex of lung

307. True about nervous system involvement in SLE is following except:
A. Seizures   PGI 1996
B. EEG abnormal in 30%
C. Elevated protein level in CSF in 50%
D. Pseudotumor cerebri   (Harrison-1923)

308. Streptococcal meningitis is usually a complication of :
AIIMS 1982
A. S.B.E.
B. Otitis media
C. Pharyngitis
D. Cellulitis   (Harrison-2480)

309. In Parkinsonism, there is :   PGI 1982
A. Intention tremor   (Harrison-154)
B. Babinski Toe sign
C. Postencephalitic form follows infection immediately
D. No loss of sensation

310. Retrobulbar neuritis is caused by :
PGI 1982, 87, 88; AMC 1984
A. Tabes dorsalis
B. Tuberculosis
C. Multiple sclerosis
D. Vitamin A deficiency
(Harrison-2453)

311. Seizures following head injury :   AIIMS 1981
A. Are inevitable
B. May indicate brain abscess
C. Usually occur immediately
D. Are more often generalized than focal   (Harrison-2358)

312. Transverse myelitis is not caused by :   PGI 1980
A. Syphilis
B. Acute bacterial infection
C. Tuberculosis
D. Rabies   (Harrison-2429)

Ans.: 305. D   306. C   307. B   308. B   309. D   310. C   311. C
312. C

313. **Commonest cause of supranuclear gaze palsy is :**   **Delhi 1982**
    A. Syphilis   B. Parkinsonism
    C. Tumors   D. Vascular lesions
                                        (Harrison-2402)
314. **Most diagnostic test of polymyositis is :**   **PGI 1996**
    A. Myoglobin in urine   B. ↑ CK   **(Harrison-2524)**
    C. EMG   D. Muscle biopsy
315. **Synkinesia is seen in :**   **PGI 1981**
    A. Trigeminal neuralgia B. Migraine   **(Harrison-3422)**
    C. Bell's palsy   D. All of the above
316. **Treatment of choice for complex partial seizure :**   **AMC 1985**
    A. Phenytoin   B. Carbamazepine
    C. Ethosuximide   D. Phenobarbitone
                                        (Harrison-2364)
317. **Todd's paralysis is a feature of :**   **AMC 1983**
    A. Grandmal epilepsy   B. Petitmal epilepsy
    C. Temporal lobe epilepsy D. Jacksonian epilepsy
318. **In Arnold Chiari malformation, there is :**   **Delhi 1985**
    A. Elongation of the modulla   **(Harrison-2430)**
    B. Herniation of cerebellar tonsils
    C. Dilatation of the central canal
    D. All of the above
319. **Dysdiadochokinesia is a sign of :**   **PGI 1982**
    A. Hysteria   B. Cerebellar ataxia
    C. Epilepsy   D. Tabes dorsalis   **(Harrison-126)**
320. **In case of Meningococcal meningitis, the organism can be removed from :**   **AIIMS 1990**
    A. Blood   B. CSF
    C. Petechiae   D. Nasopharynx
    E. All of the above   **(Harrison-2450)**
321. **Which of the following observations about narcolepsy is not true ?**
                                            PGI 1986
    A. Irresistible attacks of sleep from which patient can be aroused immediately
    B. In most cases abnormality in the region of hypothalamus can be detected
    C. Amphetamine is of value
    D. Methylphenidate is of value   **(Harrison-160)**

Ans.: 313. D   314. D   315. C   316. B   317. D   318. D   319. B
      320. E   321. B

322. **Best investigation for spontaneous subarachnoid hemorrhage is :**
    A. CT Scan   B. Angiography   **AMC 1984**
    C. Pneumoencephalogram   D. Ultrasound   **(Harrison-2389)**

323. **Lateral medullary (Wallenburg) syndrome is due to thrombosis of :**
                                                       PGI 1982
    A. Posterior inferior cerebellar artery   **(Harrison-2375)**
    B. Anterior inferior cerebellar artery
    C. Superior cerebellar artery
    D. Posterior communicating branch of middle cerebral artery

324. **Hyperpathia is seen in :**   **DNB 1988**
    A. Multiple sclerosis   B. Thalamic syndrome
    C. Brain stem lesions   D. Cerebellar lesion
                                            (Harrison-2418)

325. **There is a definite sensory level in :**   **AIIMS 1982**
    A. Subacute combined degeneration
    B. Erb's spinal syphilis
    C. Tumour of the cord
    D. Friederich's ataxia   **(Harrison-2416)**

326. **Scanning speech is seen in :**   **PGI 1984**
    A. Disseminated sclerosis   B. Parkinsonism   **(Harrison-2453)**
    C. Motor neurone disease   D. Pseudo bulbar palsy

327. **In the lower motor neurone lesion diseases, there is :**   **AIIMS 1984**
    A. Spastic paralysis
    B. There may be wasting of muscle in long standing case
    C. Exaggerated tendon reflexes
    D. Absence of faciculations   **(Harrison-2413)**

328. **Which of the following are associated with diabetic neuropathy ?**
                                                  UPSC 1984
    A. Acute mononeuropathy
    B. Asymmetric monononeuropathy multiplex
    C. Secondary Polyneuropathy
    D. All of the above   **(Harrison-2504)**

329. **A 65 year old man has had two episodes of hemiparaesthesia, diplopia and confusion which abated spontaneously. On the day of admission to the hospital he developed a full blown right aided hemiplegia. The most likely diagnosis is :**   **UPSC 1984**
    A. Cerebral haemorrhage   B. Subarchnoid haemorrhage
    C. Cerebral thrombosis   D. Cerebral embolism
                                            (Harrison-2371)

Ans.: 322. A   323. A   324. B   325. C   326. A   327. B   328. D
     329. C

330. Causes of cerebral thrombosis include :   **AMC 1986**
   A. Polycythaemia vera
   B. Sickle cell disease
   C. Dissecting aortic aneurysm
   D. Migrainous dura with persistent deficit
   E. All of the above   **(Harrison-2373)**

331. Amyotrophic lateral sclerosis is characterized by following except :
                                                TN 1989
   A. Occurs in the 5th and 6th decades   **(Harrison-2411)**
   B. Has signs of sensory loss
   C. Is symmetric
   D. Has fasciculation in upper extremities
   E. Should be differentiated from cervical spinal cord compression

332. A 40 years old male with chronic diarrhoea, ataxic gait, numbness, tingling, pigmented spots over neck has :
                                          UPSC 1989; Delhi 1987
   A. Addison's disease   B. Pellagra   **(Harrison-481)**
   C. Leukemia   D. Beri-Beri

333. Progressive multifocal leukoencephalopathy is a :   **PGI 1996**
   A. Nutritional disorder   B. Congenital disorder
   C. Slow viral disease   D. Drug induced
                                                (Harrison-2489)

334. Intention Tremor is seen in lesions of :   **Karnataka 1998**
   A. Caudate nucleus   B. Neocerebellar lesions
   C. Post central gyrus   D. Pre-frontal lobe
                                                (Harrison-125)

335. In Tabes dorsalis, unsteadiness of gait is :   **AIIMS 1983**
   A. Not seen   B. Throughout day and night
   C. More during day time   D. Less during day time
                                                (Harrison-1047)

336. General paralysis of Insane (GPI) usually follows primary infection after :   **AIIMS 1982**
   A. One year   B. Three years
   C. Five years   D. Ten years or more
                                                (Harrison-1047)

337. Neuro-arthropathy is not seen in :   **AIIMS 1983**
   A. Tabes dorsalis   B. Syringomyelia
   C. Friederich's ataxia   D. Peripheral neuritis
                                                (Harrison-2379)

Ans.: 330. E   331. B   332. B   333. C   334. B   335. D   336. D
    337. C

338. One of the following is a characteristic bed-side feature of Korsakoffs psychosis : UPSC 1996
  A. Astereognosis B. Aprtaxia (Harrison-152)
  C. Confabulation D. Hallucination

339. Duchenne's muscular dystrophy is associated with : Delhi 1986
  A. No skin abnormality B. Calves atrophy
  C. Complete heart block D. Marked lordosis
  (Harrison-2464)

340. Entire daily dose of phenytoin just once a day is now recommended because it : AIIMS 1983
  A. Sustains therapeutic plasma level
  B. Elicits better patient compliance
  C. Ensures better seizures control
  D. All of the above (Harrison-2365)

341. In "cluster headache" the patient : UPSC 1985
  A. Gets attacks of headache continuously for 2-4 weeks
  B. Remains free from attacks for 6 months to 2 years
  C. Has a typical migraine
  D. Has all of the above (Harrison-2355)

342. Pyknolepsy refers to a large number of attacks in a day of : PGI 1984

  A. Grand mal epilepsy B. Petit mal epilepsy
  C. Jacksonian epilepsy D. Infantile spasms

343. Fundus picture of stage 2 of Keith Wagener classification includes :
  A. Copper wire, AV nipping and focal spasm AIIMS 1985
  B. Silver wire, increased reflex
  C. Exudate and hemorrhagic spots
  D. Papilloedema

344. In an antibiotic treated patient the diagnosis of meningococcal meningitis is most likely to be made from : AIIMS 1986
  A. Blood culture
  B. CSF culture
  C. Latex agglutination of CSF sample for bacterial antigen
  D. Direct microscopic examination for presence of organisms
  (Harrison-2460)

345. Sudden loss of vision in a patient with diabetic retinopathy is due to: Delhi 1986
  A. Cataract B. Glaucoma
  C. Vitreous hemorrhage D. Papilloedema (Harrison-2121)

Ans.: 338. D 339. D 340. D 341. D 342. B 343. A 344. C
345. C

346. **In Tabes dorsalis the fibres affected are :**   **AMC 1982**
    A. Lateral spinothalamic tract
    B. Anterior spinothalamic tract
    C. Pyramidal fibres
    D. Dorsal column   **(Harrison-1047)**
347. **Neuropathic joints are not seen in :**   **PGI 1982**
    A. Tabes Dorsalis   B. Leprosy
    C. Diabetes mellitus   D. Myopathy   **(Harrison-2518)**
348. **A 16 year old girl after examination has gone to see a movie and suddenly finds that she cannot see. The most likely diagnosis is :**
    A. Brain tumour in optic chiasma   **UPSC 1986, 89**
    B. Disseminated sclerosis
    C. Hysterical conversion reaction
    D. Bilateral central retinal artery
349. **Hover of up knee jerk is seen in :**   **Kerala 1998**
    A. Chorea   B. Cerebral palsy
    C. Athetosis   D. None
350. **A 20-year male with idiopathic epilepsy has been controlled on phenytoin therapy for 9 years. A routine chest X-ray reveals increased interstitial and bronchovascular markings. The best course includes :**   **UPSC 1983**
    A. Inclusion of prednisone
    B. Inclusion of tridione   **(Harrison-2364)**
    C. Assessment of alveolar cell carcinoma
    D. Withdrawal of phenytoin and treatment with prednisone
351. **Most common cause of paralysis or paresis of the diaphragm due to phrenic nerve involvement is :**   **UPSC 1982**
    A. Mediastinitis   B. Tumour invasion
    C. Thyroid surgery   D. Adjacent pleural scarring
352. **Dementia associated with myoclonic jerks are the prominent findings in all of the following except :**   **Rohtak 1986**
    A. Subacute spongiform encephalopathy
    B. Unverrichi-Lundbry Lafora disease   **(Harrison-2399)**
    C. Parkinson's disease
    D. Some forms of Hallevorden
353. **Which type of Parkinsonism is most common?**   **AMC 1983**
    A. Paralysis agitan   B. Postatherosclerotic
    C. Postencephalitic   D. Phenothiazine induced
    E. Cerebral tumour   **(Harrison-2399)**

Ans.: 346. D   347. D   348. C   349. A   350. D   351. B   352. C
    353. A

354. Following are true about choreal movement, except :

AMC 1982

A. Irregular   B. Jerky
C. Present during sleep   D. Quasipurposive
E. Non-repetitive   **(Harrison-113,114)**

355. Peripheral nerve is thickened in following, except :

Delhi 1983

A. Tuberculoid Leprosy   B. Nurofibromatosis
C. Alcoholic neuropathy   D. Traumatic neuropathy
E. Familial neuropathy

356. Which one of the following is not true about upper motor neurone lesion ?   UPSC 1983, 88

A. Paralysis or weakness of individual group of muscle
B. Increased tone
C. Extensor plantar response
D. Loss of abdominal reflex
E. Decreased tendon reflexes   **(Harrison-2415)**

357. Following are true about trigeminal neuralgia, except :

Delhi 1988

A. More common in elderly people
B. Attacks of pain usually lasts for an hour or so (few sec).
C. There is sensory disturbance along the branch of trigeminal nerve
D. May provoke a spasm of facial muscles

(Harrison-2420)

358. Which one of following is untrue about Bell's palsy ?

AIIMS 1986, Delhi 1988

A. Loss of taste sensation on anterior 2/3 of tongue
B. Lower half of face is involved
C. There is oedema in 3-4 weeks
D. Recovery starts in 3-4 weeks
E. Saliva runs from angle of mouth   **(Harrison-2421)**

359. Hitzig zones are seen in :   TN 1998

A. Neurosyphilis
B. Tabes dorsalis
C. Amyotrophic lateral sclerosis
D. Multiple sclerosis

360. "Button Hole' Sign is present in :   Rajasthan 1994

A. Angiofibroma   B. Neuroma
C. Neurofibroma   D. Adnexal tumours

Ans.: 354. C   355. E   356. E   357. B   358. A   359. B   360. C

361. **The classical diagnostic triad of mental deterioration, visual failure and paralysis is seen in :**   **DNB 1991**
   A. Hand-Schuller-Christian syndrome
   B. Niemann-Pick disease
   C. TaySach's disease
   D. Gaucher's disease   **(Harrison-2413)**

362. **Drug of choice in West syndrome (mental retardation with salaam fits) is :**   **AIIMS 1985**
   A. Carbamazepine   B. Clonazepam
   C. Sodium valproate   D. ACTH

363. **In some patients, although walking is so difficult, they can be seen running for a bus quite well (kinesia Paradoxa) is typically seen in :**   **PGI 1982**
   A. Wilson's disease   B. Cerebellar tumor
   C. Parkinsonism   D. Astasiaabasia (hysteria)
                    (Harrison-2399)

364. **Postencephaltic type of Parkinsonism differs from idiopathic paralysis agitans that :**   **DNB 1989**
   A. Tremors always precede rigidity
   B. Rigidity always precede tremors
   C. Both appear simultaneously
   D. None of the above   **(Harrison-2399)**

365. **'Eosinophilic meningitis' results from :**   **Delhi 1998**
   A. Toxoplasma gondii   B. Toxocara canis
   C. Oesophagostomum   D. Angiostrongylus

366. **Most useful in the treatment of Sydenham chorea (rheumatic chorea) is :**   **DNB 1988**
   A. Penicillin
   B. Aspirin
   C. Bed rest followed by gradual mobilisation
   D. Chlorpromazine   **(Harrison-125)**

367. **Which of the following symptoms of Parkinsonism responds least to levodopa :**   **AIIMS 1987**
   A. Tremor   B. Rigidity
   C. Akinesia   D. Gait   **(Harrison-2400)**

368. **Sanger Brown's ataxia differs from Friedreich's ataxia by the following except :**   **PGI 1987**
   A. Late age of onset
   B. Occurrence of ptosis, optic atrophy and ophthalmoplegia

---

Ans.: 361. C   362. D   363. C   364. B   365. D   366. C   367. D
   368. D

C. Spasticity in lower limbs
D. Nystagmu

369. A seizure arising in one motor cortex starts most frequently in any of the following except :   AP 1987
A. Thumb
B. Eyelid
C. Angle of mouth
D. Great toe   (Harrison-2355)

370. Most common type of meningoencephalocele is :
TN 1988
A. Frontal
B. Temporal
C. Parietal
D. Occipital

371. Distinctive EEG showing bursts of triphasic slow waves is often seen in :   AMU 1985
A. Benign lymphocytic meningitis
B. Hypoglycemia
C. Postictal phase
D. Subacute sclerosing panencephalitis   (Harrison-2362)

372. Spironolactone bodies are seen in :   Assam 1995
A. Neurone
B. Mitochondria
C. Hippocampus
D. None of the above

373. Migraine may be associated with all of the following except :
AMC 1985
A. Paraesthesia
B. Dysphagia
C. Seizures
D. Diplopia   (Harrison-2351)

374. Which of the following supports the correct diagnosis of paralysis agitans :   PGI 1987
A. Oculogyric crises
B. Slow monotonous speech
C. Loss of emotional control
D. Bilateral extensor plantar   (Harrison-2399)

375. All are true about neurosyphilis except :   AIIMS 1985
A. Intramuscular procaine benzylpenicillin is the treatment of choice
B. Near all patients with active disease show high cell counts in the CSF
C. The raised CSF cell count comes to normal as a result of effective therapy
D. Treatment of neurosyphilis should be continued till the CSF Wassermann reaction is negative
(Harrison-1047)

Ans.: 369. B   370. D   371. D   372. C   373. C   374. B   375. D

376. **The phenomenon of "sparing of the macula" is due to the collateral circulation between :**   **Karnataka 1995**
    A. Middle and posterior cerebral arteries
    B. Anterior and middle cerebral arteries
    C. Anterior and posterior cerebral arteries
    D. Anterior, middle and posterior cerebral arteries

377. **The common type of cerebral palsy seen in hospital is :**
                                                C.U.P.G.E.E. 1995
    A. Spastic   B. Monoplegia
    C. Quadriplegia   D. Diplegia

378. **In CSF all the following ions are in equal concentration to that of plasma except :**   **TN 1989**
    A. Sodium   B. Potassium
    C. Chloride   D. Urea   **(Harrison-2444)**

379. **Akinetic mutism is seen in all of the following except :**
                                                AIIMS 1983
    A. Bilateral frontal lobe lesions B. Diffuse cortical damage
    C. Cerebellar lesion   D. Mid-brain lesion
                                                (Harrison-136)

380. **Which of the following is not associated with von Recklinghausen's disease ?**   **AIIMS 1984**
    A. Optic glioma   B. Phaeochromocytoma
    C. Diabetes mellitus   D. Acoustic neuroma
    E. Kyphoscoliosis   **(Harrison-2445)**

381. **Anticoagulation is indicated in neurologic deficit due to all of the following except :**   **Karnataka 1996**
    A. Mitral stenosis with atrial fibrillation
    B. Postpartum sagittal sinus thrombosis
    C. Infective endocarditis
    D. Stuttering hemiplegia

382. **Triad diagnostic of pin-point pupil, pyrexia, paralysis is :**
                                         AIIMS 1983, UPSC 1983
    A. Morphine poisoning   B. Subdural hemorrhage
    C. Pontine hemorrhage   D. Midbrain lesion
                                (Harrison-2387)

383. **Hydrocephalus is associated with :**   **Delhi 1998**
    A. Hyperaemic optic disc   B. Macular edema
    C. Unilateral optic atorphy D. Bilateral optic atrophy

Ans.: 376. E   377. D   378. C   379. D   380. C   381. C   382. C
    383. D

384. Athetosis is most often found in association with : AMU 1990
   A. Disseminated sclerosis   B. Parkinson's disease
   C. Huntington's chorea   D. Sydenham's chorea
   E. Cerebral palsy   (Harrison-125)

385. All of the following are used in the treatment of Gullain Barre syndrome, except : Karnataka 1996
   A. Plasmapheresis
   B. I.V. methyl prednisolone
   C. I.V. gammaglobulin
   D. Large doses of neuro vitamins

386. Bilateral sixth cranial nerve palsies may occur in the following except: DNB 1989
   A. Tuberculous meningitis   B. Nasopharyngeal carcinoma
   C. Guillain-Barre syndrome   D. Raised intracranial pressure
   E. Motor neurone disease

387. Bilateral facial wakness is a feature of following except :
   A. The Guillain-Barre syndrome   Delhi 1983; JIPMER 1993
   B. Myasthenia gravis
   C. Pontine glioma
   D. Multiple sclerosis
   E. Sarcoidosis   (Harrison-2422)

388. Characteristic findings in the CSF of a patient with multiple sclerosis include : UPSC 1987
   A. Positive syphitic serology
   B. Oligoclonal pattern of immunoglobulins on electrophoresis
   C. Total protein of 1.2 g/l (120mg/100 ml)
   D. Increased white cell count especially polymorphs
   E. Xanthochromia   (Harrison-2454)

389. Most common cranial nerve involved in ophthalmoplegic migraine: AP 1996
   A. II nerve   B. III nerve
   C. V nerve   D. VI nerve

390. The phenomenon of "sensory inattention" can be found in patients with lesions of : Bihar 1989
   A. Thalamus
   B. Betz area
   C. Brain stem and spinal cord
   D. Frontal lobe of cerebral cortex
   E. Parietal lobe of cerebral cortex   (Harrison-135)

Ans.: 384. E   385. B   386. E   387. D   388. B   389. B   390. E

391. In which type of parkinsonism maximum degree of rigidity is seen :
                                                        AMC 1984, Bihar 1991
    A. Postencephalitic   B. Paralysis agitans
    C. Drug induced   D. All of the above
    E. Both A +C above   **(Harrison-2399)**

392. Which of the following is not true about acromegaly ?
                                                        AMU 1986
    A. Homonymous hemianopia is the most important visual complication
    B. Characteristic enlargement of the frontal sinuses
    C. Galactorrhoea may be a presenting symptoms
    D. Increased resistance to insulin
    E. Painful in the distribution of median nerve paraesthesia
                                                        (Harrison-2045)

393. Cerebral embolism occurs usually due to impactation of the embolus into :   **Delhi 1983**
    A. Middle cerebral arteries   B. Posterior cerebral artery
    C. Vertebral artery   D. Internal carotid artery
                                                        (Harrison-2412)

394. Rapid onset of oculomotor palsy associated with pain around the eye may suggest all of the following except :   **Rohtak 1987**
    A. Myasthenia gravis   B. Migraine
    C. Intracranial aneurysm   D. Diabetic mononeuropathy
                                                        (Harrison-2510)

395. A patient with ptosis, generalized weakness and difficulty is swallowing has :   **AIIMS 1992**
    A. Familial periodic paralysis   B. Myasthenia gravis
    C. Myotonia dystrophica   D. Thyrotoxic myopathy
                                                        (Harrison-2511)

396. Basal ganglia lesions are seen in: **Delhi 1998**
    A. Alzheimer's disease
    B. Creutzfeldt Jacob's disease
    C. Neurolathyrism
    D. Sub acute sclerosing panencephalitis   **(Harrison-2491)**

397. In syringomyelia, the sensation that is lost is :
                                                        AIIMS 1986; TMC 1985
    A. Touch   B. Pressure
    C. Pain and temperature   D. Two point discrimination
                                                        (Harrison-2430)

Ans. 391. B   392. A   393. A   394. A   395. B   396. B   397. C

398. The highest frequency wave in EEG is ——— wave :

A. Alpha   B. Beta   **Delhi 1986**

C. Delta   D. Theta

399. Parkinsonism like rigidity is caused by all except :

                    PGI 1984; AIIMS 1988; AI 1997

A. Phenothiazines

B. CO poisoning

C. Vertebrobasiliar insufficiency

D. Metoclopramide intake   **(Harrison-2399)**

400. A patient presents with loss of touch and temperature sensation on one side of face with loss of sensation on other side of the body. The most likely site of lesion is :   **Delhi 1997**

A. Cerebral cortex   B. Internal capsule

C. Pons   D. Medulla

401. Muscle not usually involved in polymyositis are :

                    AIIMS 1990

A. Limb musles   B. Ocular muscles

C. Pharyngeal muscles   D. Jaw muscles

                    (Harrison-2524)

402. Earliest symptom of myaesthenia is :   **AIIMS 1988**

A. Diplopia + ptosis   B. Fatigue

C. Respiratory weakness   D. Palatal palsy

403. Alcohol is used in Rx of :   **Kerala 1997**

A. Trigeminal neuralgia   B. Status epilepticus

C. Chronic insomnia   D. Abreaction

404. Which of the following is not involved in lateral medullary syndrome:

                    AIIMS 1982

A. Sympathetic tract   B. IX, X, XI cranial nerve

C. XII cranial nerve   D. Spinothalamic tract

                    (Harrison-2375)

405. Commonest cause of chorea is :   **AMC 1990**

A. Huntiugton's disease   B. Levodopa toxicity

C. Alcoholism   D. Rheumatic fever

                    (Harrison-2341)

406. A patient can not move his limb against resistance offered by examiner but can move limb against gravity. What is the grade of power :   **JIPMER 1997**

A. I   B. II

C. III   D. IV

Ans.: 398. B   399. C   400. D   401. B   402. A   403. A   404. C

      405. D   406. C

407. Wilson's disease is characterised by following except :   AIIMS 1990
  A. Spasticity   B. Sensory impairment
  C. Rigidity   D. Chorea   **(Harrison-2274)**

408. Drug of choice in drug induced parkinsonism is :   AI 1990
  A. Levodopa   B. Benzhexol   **(Harrison-2399)**
  C. Amantidine   D. Carbidopa

409. False about CNS lymphoma is :   AIIMS 1997
  A. Not responsive to radiotherapy and chemotherapy
  B. Seen in HIV
  C. Multicentric and CSF spread
  D. Diffuse histiocytic variant   **(Harrison-2493)**

410. The conversion of extensor plantar response in infancy to flexor plantar type occurs usually in the age group of :   DNB 1990
  A. 3-6 months   B. 6-9 months
  C. 9-12 months   D. 20-24 months

W411. Which of the following disorder is associated with early onset of Alzheimer's disease :   Delhi 1998
  A. Turner syndrome   B. Huntington's disease
  C. Down syndrome   D. Hypothyroidism
                                    (Harrison-2392)

W412. Spasticity of right lower limb is caused due to injury to tract :
                                       Orissa 1998
  A. Right spinothalamic   B. Right corticospinal
  C. Right spinocerebellar   D. Left dorsal column

413. Earliest sensation lost in diabetic neuropathy :   AIIMS 1997
  A. Pain   **(Harrison-2122)**
  B. Temperature
  C. Vibration
  D. Weakness of small muscles of hand

414. Commonest neoplastic cause of intracranial calcifi-cation in a child is :   Delhi 1988, 90
  A. Craniopharyngioma   B. Medulloblastoma
  C. Meningioma   D. Ependymoma
                                    (Harrison-2034)

415. Transverse myelitis usually affects :   PGI 1987
  A. Thoracic and lumbar segments
  B. Cevical and thoracic segment
  C. Lumbar and sacral segments
  D. Lumbar segment only   **(Harrison-2429)**

Ans.: 407. B   408. B   409. A   410. C   411. A, D   412. A, B, C
        413. C   414. A   415. A

416. Persistent contractions of muscle despite voluntary attempts at relaxation are seen in all of the following, except : UPSC 1987
A. Myotonia dystrophica
B. Myasthenia gravis
C. Myotonia congenita
D. Normokalaemic periodic paralysis (Harrison-2515)

417. Contralateral upper trunk paralysis is due to thrombosis of : AIIMS 1997
A. Ant. cerebral artery
B. Middle cerebral artery
C. Ant. choroidal artery
D. Post. choroidal artery
(Harrison-2382)

418. Which of the following is a cause of bilateral hypoglossal nerve paralysis : DNB 1990
A. Syringobulbia
B. Head injury
C. Aneurysm of verterbral artery
D. Progressive bulbar paralysis (Harrison-2423)

419. Acute attack of migraine is best abolished by : AIIMS 1997
A. Ergotamine
B. Sumitriptan
C. Propranolol
D. Amitriptyline
(Harrison-2352)

420. Neuromuscular claudication is due to : AIIMS 1997
A. Syringomyelia
B. Osteoarthritis
C. Lumbar canal stenosis
D. Buerger's disease
(Harrison-82)

421. Following causes of Argyll-Robertson pupil except : DNB 1990
A. Syringomyelia
B. Disseminated sclerosis
C. Chronic alcoholism
D. Asphyxia and deep anaesthesia (Harrison-165, 1048)

422. Bilateral pyramidal signs develop early in : PGI 1989
A. Cerebral hemorrhage
B. Subarachnoid hemmorrhage
C. Pontine hemorrhage
D. Hematomyelis (Harrison-2428)

423. Most frequent finding of Bell's palsy is : PGI 1989
A. Diminished corneal reflex
B. Miotic pupil
C. Absence of taste sensations
D. Anhidrosis over half of face

Ans.: 416. B   417. B   418. D   419. B   420. C   421. D   422. D
423. A

424. **Bensarazide is useful in parkinsonism because it is a :**   **PGI 1989**
   A. Facilitator of dopaminergic transmission like amantadine
   B. Peripheral dacarboxylase inhibitor
   C. Centrally acting anticholinergic drug like benzhexol
   D. Dopaminergic agonist like bromocriptine
                                                    (Harrison-2401)
425. **Waddling gait is encountered in the following except :**   **PGI 1989**
   A. Pseudohypertrophic muscular dystrophy
   B. Osteomalacia
   C. Congenital dislocation of hip
   D. Friederich's ataxia   **(Harrison-127)**
426. **Bilateral symmetrical brisk reflexes may be seen in :**   **AMU 1988**
   A. Pyramidal tract lesion   B. Hystseria
   C. Neurasthenia   D. All of the above
427. **Best diagnostic aid to myasthenia gravis is :**   **AIIMS 1987**
   A. History
   B. EMG
   C. Chest X-ray
   D. Response to IV edrophonium   **(Harrison-2516)**
428. **Inverted supinator jerk indicates lesion at :**
                                              AIIMS 1983; AI 1996
   A. $C_{4,5}$ B. $C_{5,6}$
   C. $C_{7,8}$ D. C_8T_1
429. **Extensor plantar response will not be detected even in the presence of a pyramidal lesion if there is presence of all of the following except:**
                                                    AIIMS 1985
   A. Gross loss of senation on the sole of the foot
   B. Hallux rigidus
   C. Paralysis of extensor hallucis longus
   D. Paralysis of extensor hallucis brevis
430. **"Locked-in-syndrome" may be seen in except :**   **DNB 1990**
   A. Pontine lesions
   B. Tegmentum lesions
   C. Bilateral lesions of medulla
   D. Bilateral lesions of lateral one-third of cerebellar peduncle
                                                    (Harrison-126)
431. **The cause of slowly progressive paraplegia, root pains and patchy sensory loss is :**   **PGI 1982; AIIMS 1984**
   A. Motor neurone disease   B. Fluorosis
   C. Leprosy   D. Cysticercosis

Ans.: 424. B   425. D   426. D   427. D   428. B   429. D   430. D
      431. B

432. **Which of the following is not a feature of midbrain lesion :**
    AIIMS 1984, 87
    A. Upper gaze palsy   B. Pupillary dilatation
    C. Crossed hemiplegia   D. Extension of arm
                                          (Harrison-2376)

433. **Pure motor neuropathy is seen in following except :**   AIIMS 1997
    A. Head toxicity   B. Porphyria
    C. Guillain Barre Syndrome D. None of the above
                                     (Harrison-640, 2498)

434. **L-dopa is useful in Parkinsonism because :**   AIIMS 1984, 85
    A. It is converted to Norepinephrine
    B. It is converted to a dopamine precursor in blood
    C. It is converted to dopamine in brain
    D. It downregulates the cholinergic activity   **(Harrison-2401)**

435. **The largest branch of internal carotid artery the one with the highest risk for emboli to pass into is :**   Kerala 1998
    A. Middle cerebral
    B. Ant. cerebral
    C. Posterior communicating
    D. Posterior cerebral   **(Harrison-2383)**

436. **Biological amnesia is due to :**   AIIMS 1998
    A. Senile dementia   B. Lack of interest
    C. Opioid poisoning   D. Anxiety neurosis

437. **Hypotonia, tremor, ataxia are seen in lesion of :**   PGI 1998
    A. Basal ganglion   B. Medullary pyramid
    C. Cerebellum   D. Pons   **(Harrison-126)**

438. **Reticular activating system has following functions :**   PGI 1989
    A. Wakefulness   B. Memory
    C. Posture   D. Sleep cycle
    E. Coordination   **(Harrison-136)**

439. **Prolonged ankle jerks are seen in following except :**   AMU 1986
    A. Myasthenia gravis   B. Parkinsonism
    C. Hypokalemia   D. Hyperthyroidism

440. **Unilateral past pointing nystagmus is seen in :**   PGI 1986
    A. Superior semicircular canal involvement
    B. Posterior semicircular canal involvement
    C. Cerebellar vermix lesion
    D. Flocconodular lesion in cerebellum
    E. None of the above   **(Harrison-182)**

Ans.: 432. D   433. C   434. C   435. A   436. A   437. C   438. A
       439. D   440. D

441. The vitamin contraindicated in therapy with levodopa is :
                                                            PGI 1982
   A. B2   B. Pyridoxine
   C. Folic acid   D. Ascorbic acid
                                                            (Harrison-2400)

442. Which of the following is most important useful sign in left temporal lobe abscess :   **AIIMS 1985, 87**
   A. Deep left temporal headache
   B. Seizures
   C. Vomiting
   D. Nominal aphasia   **(Harrison-153)**

443. Guillan Barre Syndrome, followng are seen except :   AIIMS 1997
   A. CSF pleocytosis
   B. Areflexia
   C. Motor loss with paresthesia
   D. Ascending paralysis

444. Botulin toxin is used in :   **PGI 1998**
   A. Focal dystonia   B. Myaesthenia gravis
   C. Cerebellar ataxia   D. Hypotonia   **(Harrison-920)**

445. Peripheral neuropathy is caused due to :   **AIIMS 1997**
   A. Bleomycin   B. Busulphan
   C. Cisplatin   D. Cyclophosphamide
                                                            (Harrison-2498)

446. Normal CSF pressure is :   **UPSC 1984; AIIMS 1986**

	mmCSF	*mmHg*	*N/m2*
A.	60-120	1-3	200-400
B.	120-180	3-12	400-1600
C.	180-250	12-15	1600-1800
D.	250-400	15-20	1830-2200

447. In DM neuropathy not used is :   **AIIMS 1997**
   A. Amitriptyline   B. Local Capsium
   C. Dextroamphetamine   D. Phenytoin   **(Harrison-2506)**

448. Most characteristic CNS lesion of HIV is :   **AIIMS 1997**
   A. Perivascular Giant cells   B. Vasculitis
   C. Spongiform degeneration   D. Microglial nodule

449. Spinomuscular atrophy is seen in disease of :   **PGI 1998**
   A. Ant. horn   B. Peripheral nerve
   C. Neuromuscular junction   D. Any of the above
                                                            (Harrison-119)

Ans.: 441. B   442. D   443. A   444. A   445. C   446. B   447. C
      448. D   449. A

450. **True about inferior Frontal Gyrus lesions on domi-nant side is :**
     AIIMS 1997
     A. Compromised speech
     B. Defective comprehension of written language
     C. Dysarthria
     D. Defective comprehension of hearing
451. **Negri bodies are most often found in :** AIIMS 1985; UPSC 1991
     A. Midbrain   B. Basal ganglia
     C. Frontal cortex   D. Hippocampus
                                       (Harrison-1149)
452. **In Korasakoff's psychosis, lesion is present at :** **AIIMS 1985**
     A. Uncinate body   B. Mamillary body
     C. Cingular gryus   D. Frontal lobe
453. **Ataxic nystagmus indicates a lesion in :** **JIPMER 1992**
     A. Medial longitudinal fasciculus
     B. Labyrinth
     C. Vestibule
     D. Cerebellum   **(Harrison-182)**
454. **Equine gait is a feature of :** **JIPMER 1992**
     A. Extrapyramidal dysfunction B. Brainstem lesion
     C. Anterior tibial N. injury   D. Multiple sclerosis
455. **Mask like face is seen in :** **PGI 1986**
     A. Parkinsonism   B. Disseminated sclerosis
     C. After stroke   D. Pseudo bulbar palsy
                                       (Harrison-2399)
456. **Presence of hemiplegia with diminution of vision in the contralateral**
     eye suggests occlusion of : **UP 1997**
     A. Middle cerebral artery   B. Basilar artery
     C. Anterior cerebral artery D. Internal carotid artery
                                       (Harrison-2382)
457. **Intentional tremors in young patients is commonly due to :**
     Delhi 1984
     A. Progeria   B. Wilson's disease
     C. Parkinsonism   D. Head injuries **(Harrison-123)**
458. **Following are early CNS complications of AIDS except :**
     A. CNS lymphoma   **PGI 1993**
     B. Aseptic meningitis
     C. Dementia
     D. Demyelinating encephalopathy   **(Harrison-1891)**

Ans.: 450. A   451. D   452. B   453. A   454. C   455. A   456. A
       457. B   458. A

459. The most common cranial nerves involved in leprosy (neuritic type) are :   PGI 1985, 87
    A. V, VI   B. V, VII
    C. VI, VII   D. IV, V
    E. VI, VII   (Harrison-2508)

460. Downbeat nystagmus is characteristic of :   DNB 1992
    A. Posterior fossa lesions   B. Vestibular lesions
    C. Labyrinthine lesions   D. Cerebellar lesions
                                                        (Harrison-182)

461. Most common cause of stroke in young women in India is :
                                                          PGI 1998
    A. Central vein thrombosis   B. Head injury
    C. Atherosclerosis   D. HT

462. Following is characteristic neurologic finding in primary amyloidosis: AMC 1984; AIIMS 1986
    A. Peripheral motor and sensory neuropathy
    B. Peripheral neuropathy associated with cerebral manifestation
    C. Guillain-Barre type of syndrome
    D. Spinal cord compression in thoracic region   (Harrison-1974)

463. In primary idiopathic polymyositis, the following group of muscles is almost never affected :   UP 1997
    A. Proximal limb girdle muscles
    B. Pharyngeal muscles
    C. Extensor neck muscles
    D. Ocular muscles   (Harrison-2524)

464. Steele-Richardson syndrome is characterized by following except :
                                                          PGI 1997
    A. Convulsions   B. Ataxia
    C. Nystagmus   D. Dementia   (Harrison-2402)

465. An 8 year old girl is admitted with low grade fever, headache, vomiting and attacks of convulsions. On local examination no localising sign is demonstrated. The CSF examination reveals protein 150 mg%, sugar 40 mg% chloride 520mg% and cell count of 100/cu min with large number of lymphocytes. Most likely diagnosis is :
                                                         AIIMS 1986
    A. Pyogenic meningitis   B. Tuberculous meningitis
    C. Cerebral abscess   D. Viral encephalitis
                                                        (Harrison-2477)

Ans.: 459. B   460. A   461. A   462. A   463. D   464. A   465. B

466. **Ankle jerk is lost in :**   PGI 1997
    A. Trauma to knee   B. Osteoarthritis
    C. Severe anemia   D. Elderly
467. **A lesion of the oculomotor nerve can result in :**   UPSC 1987
    A. Deviation of eye inward
    B. Ptosis of the eyelid
    C. Widening of palpebral fissure
    D. All of the above   (Harrison-180)
468. **Alzheimer's disease, which of the following neurotransmitter is mainly involved :**   DNB 1992
    A. GABA   B. 5-HT   (Harrison-2392)
    C. Dopamine   D. Acetylcholine
469. **The treatment of choice in carcinomatous myaesthenic syndrome :**
                                                      AIIMS 1981, 83
    A. Steroids   B. Guanidine   (Harrison-2414)
    C. Indomethacin   D. Radiotherapy
470. **Allodynia is defined as :**   AIIMS 1998
    A. Painful perception for usually non-painful stimuli
    B. ↑ Pain sensation for painful stimuli
    C. Non painfulness for painful stimuli
    D. ↓ pain sensation   (Harrison-2418)
471. **A raised serum creatine phosphokinase level is ususual in :**
    A. Acute alcoholic myopathy   UPSC 1997
    B. Viral polymyositis
    C. Myopathy of Cushing's syndrome
    D. Duchenne's muscular dystrophy   (Harrison-2091)
472. **HLA associated with myasthenia gravis is :** PGI 1984, 89
    A. B_{27} B. D W_3
    C. DR_4 D. B_8 (Harrison-2510)
473. **Ligation of the internal carotid artery results in death and hemiplagia in ——% of cases :**
    A. 33   B. 50
    C. 75   D. 90
    E. 100
474. **Among the cranial nerve palsies the following is common in diphtheria :**   Karnataka 1987
    A. Paralysis of facial nerve   B. Ptosis
    C. Paralysic of soft palate   D. Optic atrophy (Harrison-910)

Ans.: 466. D   467. B   468. D   469. B   470. A   471. C   472. D
       473. A   474. C

475. In Fredereich ataxia, earliest presentation :   **AIIMS 1993**
    A. Ataxia   B. Stuttering
    C. Optic atrophy   D. Seizures   **(Harrison-2409)**

476. In patient of AIDS, chorioretinitis is caused by :   **AIIMS 1993**
    A. Toxoplasma gondii   B. Cytomegalovirus
    C. Candida   D. Herpes simplex
                                        (Harrison-1882)

477. Rhinocerebral mucormycosis is a complication in following except:
                                        AIIMS 1993
    A. Uncontrolled diabetes mellitus
    B. CRF
    C. Desferroxamine therapy
    D. Organ transplantation   **(Harrison-1179)**

478. Autonomic Neuropathy commonly occurs in all of the following except: **Karnataka 1994**
    A. Leprosy   B. Guillain Barre Syndrome
    C. Lead polyneuropathy   D. Porphyria   **(Harrison-2508)**

479. Flapping tremors are seen in except :
                   AI 1989; TN 1989; MAHE 1999
    A. Thyrotoxicosis   B. CO_2 narcosis
    C. Uraemia   D. Hepatic failure

Directions : The next two items are based on the following case history. Study the same carefully and attempt the two items that follow it. A 16-year old girl complains of progressive difficulty in climbing stairs and pain in the legs of two months duration. On examination, she has weakness of neck muscles and proximal muscles of upper and lower limbs. There is mild tenderness of muscles. Deep tendon reflexes are normal.

480. The most likely diagnosis is :   **UPSC 1997**
    A. Polymyositis   B. Poliomyelitis
    C. Muscular dystrophy   D. Guillain-Barre syndrome
                                       (Harrison-2514)

481. Which of the following investigation will clinch the diagnosis in this case ?   **UPSC 1997**
    A. Cerebrospinal fluid examination
    B. Muscle biopsy
    C. Nerve conduction studies
    D. Gulillain-Barre syndrome   **(Harrison-2515)**

Ans.: 475. A   476. B   477. B   478. B   479. A   480. C   481. B

482. **A 45 years old male has an acute attack of vertigo, vomiting and ataxia. The most likely diagnosis is thrombosis of :**   **AI 1994**
   A. Posterior inferior cerebellar artery
   B. Superior cerebellar artery
   C. Middle cerebellar artery
   D. Posterior cerebellar artery

483. **A young man has suddenly developed paraplegia following a short febrile illness. Examination reveals lower motor neurone type paralysis with no sensory disturbances and no bladder and bowel involvement. The C.S.F examination in this case of likely to reveal :**   **UPSC 1994**
   A. Normal C.S.F.
   B. Traces of blood
   C. Albumino-cytological dissociation
   D. Features of tuberculous meningitis

484. **Which of the following is not a slow virus disease :**   **UPSC 1994**
   A. Progressive multifocal leucoenphalopathy
   B. Progressive rubella encephalitis
   C. Limbic encephalitis
   D. Kuru   **(Harrison-2459)**

485. **Radioisotopic brain scan is used chiefly to localize :**   **PGI 1987**
   A. Brain abscess   B. Subdural haemtomas
   C. Supratentorial tumors   D. Infratentorial tumours

486. **Features of Fredrickson's Type IV hyperlipo-proteinemia include all of the following except :**   **UPSC 1994**
   A. Cholesterol levels generally normal
   B. Clear plasma
   C. Accelerated coronary artery disease
   D. Glucose intolerance   **(Harrison-2410)**

487. **Selegilline hydrochloride is a useful drug in the treatment of :**   **UPSC 1994**
   A. Migraine   B. Parkinsonism
   C. Cerebral thrombosis   D. Subarachnoid hemorrhage
   (Harrison-2400)

488. **Sodium valproate alters the levels of neurotransmitter :**   **DNB 1993**
   A. Dopamine   B. Norepinephrine
   C. 5-HT   D. GABA   **(Harrison-2364)**

Ans.: 482. A   483. C   484. B   485. C   486. B   487. B   488. D

489. Pyramidal lesions are characterized by all of the following except :
                                                                    AI 1994
A. ↑ Tendon reflexes   B. Babinski sign present
C. Involuntary movements   D. Clasp knife rigidity

490. Macular vision and adult acquity is attained by a   child at the age (years) of — respectively :   **DNB 1991**
A. 3, 5   B. 4, 6
C. 5, 7   D. 6, 8

491. True about migraine is :   **DNB 1994**
A. Usually starts before puberty
B. Attack decreases with age
C. Continue till oldage
D. Flunarizine is useful in acute attack   **(Harrison-73 to 76)**

492. Dementia can occur in which of the following cases ?   UPSC 1997
1. Sub-acute spongiform encephalopathy
2. Human immunodeficiency virus infection
3. Cerebral arteriovenous malformation
4. Prion disease
Select the correct answer using the codes given below :
Codes :
A. 1, 2 and 3   B. 2 and 4
C. 1,2 and 4   D. 3 and 4   **(Harrison-2459)**

493. Leucoencephalopathy is caused by :   **PGI 1994**
A. Vincristine   B. Cisplatinum
C. Methotrexate   D. Adriamycin   **(Harrison-2489)**

494. Which of the following diseases may be associated with Guillain Barre syndrome :   **AMU 1987**
A. Infectious mononucleosis   B. Rubeola
C. Mumps   D. Scarlet fever
E. All of the above   **(Harrison-2415)**

495. Which of the following reduces the intracranial pressure :
                                                            PGI 1994
A. Hypertonicity of plasma   B. Hypotnicity of plasma
C. Hypernatraemia   D. Hypercapnia   **(Harrison-2438)**

496. A 25 year old male patient presents with severe headache followed by unconsciousness CSF tap revealed red blood. Most likely diagnosis: **AIIMS 1994**
A. Subdural haematoma   B. Ruptured aneurysm
C. Embolism   D. Tumour   **(Harrison-2386)**

Ans.: 489. C   490. C   491. B   492. C   493. C   494. E   495. B
    496. B

497. Which is not a feature of Sturge Weber syndrome :   AIIMS 1994
   A. Snail track calcifications   B. Hemiatrophy of brain
   C. Convulsions   D. Empty sella
                                      (Harrison-2445)

498. Injection of hypertonic Saline in which region of hypothalamus produces intense thirst :   AIIMS 1994
   A. Paraventricular   B. Supraoptic
   C. Preoptic   D. Posterior region
                                      (Harrison-2035)

499. Embolism of cerebral artery leads to memory impairment because of damage to :   AIIMS 1994
   A. Hippocampal gyrus   B. Angular gyrus
   C. Premarginal area   D. Superior temporal gyrus
                                      (Harrison-2393)

500. Extensive lesion of somatic sensory area of cortex lead to impairment in following except :   AIIMS 1994
   A. Sensory localization   B. Extensive pressure
   C. Exact weight determination   D. Pain

501. Autonomic neuropathy in Diabetes mellitus is often characterized by following except :   AIIMS 1996; AI 1997
   A. Hypertension   B. Diarrhoea or constipation
   C. Impotence in males   D. Difficulty in swallowing
                                      (Harrison-2123)

502. III nerve palsy with pupillary sparing is seen in :   AIIMS 1995
   A. DM
   B. Aneurysm of post communicating artery
   C. Hypertension
   D. Craniopharyngioma   **(Harrison-2123)**

503. Causes of acute ascending motor paralysis are following except :   AI 1995
   A. Diphtheria   B. Diabetes mellitus
   C. Guillain barre syndrome   D. IMN   **(Harrison-2432)**

504. Disorder of language of cerebral origin is :   AI 1995
   A. Aphasia   B. Dysarthria
   C. Sluttering   D. Echolalia   **(Harrison-140)**

505. Horner's syndrome is produced due to pressure on :   AI 1995
   A. Stellate ganglia   B. Branchial pelxus
   C. Coeliac plexus   D. Cervical plexus
                                      (Harrison-165, 178)

Ans.: 497. D   498. B   499. A   500. D   501. A   502. A   503. B
   504. A   505. A

506. A 30 year old woman presents with a history of amenorrhoea and impaired vision of six months duration. Physical examination shows everything normal except for pale optic discs and diminished visual acuity. The most likely diagnosis is :   UPSC 1995
    A. Pituitary adenoma
    B. Craniopharyngioma
    C. Hypothalamic glioma
    D. Benign intracranial hypertension   (Harrison-2034)

507. A 40 years old female presents with a history of sudden onset of headache and nausea which passed off with rest and anaglesics. Later she developed blurring of vision for a few days. On the day of admission she had developed third nerve palsy with neck rigidity. The temperature was 100 degree F. The most likely diagnosis is :   UPSC 1995
    A. Acute encephalitis   B. Viral encephalitis
    C. Sub-arachnoid haemorrhage   D. Severe hypertension
                                              (Harrison-2388)

508. A 40 year old male, who is known to the HIV positive has been admitted with a history of fever, headache, vomiting and drowsiness for the seven days. Which of the following investigations would be helpful in establishing the etiological diagnosis ?   UPSC 1995
    1. CT of head
    2. CSF serology
    3. India ink preparation of CSF   (Harrison-1892)
    Select the correct answer using the codes given below :
    A. 1 and 3   B. 2 and 3
    C. 1 and 2   D. 1,2 and 3

509. Presence of calcification in an intracranial lesion is best made out by :   UPSC 1995
    A. CT   B. MRI
    C. Ultrasound   D. Contrast study

510. 'Hyrano bodies' are seen in :   AP 1993
    A. Alzheimer's disease   B. Huntington's disease
    C. Parkinson's disease   D. Pick's disease

511. C.S.F. seedling occurs in :   AP 1993; PGI 1994
    A. Meningioma   B. Medulloblastoma
    C. Oligodendroglioma   D. Pineal body tumours
                                              (Harrison-2443)

Ans.: 506. B   507. C   508. D   509. A   510. A   511. B

512. **The following are true in a case of brain stem glioma except :**
    AP 1993
    A. Biopsy is mandatory
    B. Surgery is difficult
    C. M.R.I. delineates tumours well
    D. Radiotherapy is the treatment of choice

513. **Neurofibromatosis is not commonly associated with :** **AP 1993**
    A. Meningocele
    B. Meningioma
    C. Pigmented spots of skin
    D. Cerebral artery aneurysm
    (Harrison-596, 244?)

514. **The least common site for the origin of meningioma is :**
    AP 1993
    A. Falx
    B. Tentorium
    C. Sphenoid wing
    D. Choroid plexus
    (Harrison-2444)

515. **Mode of action of Bromergocriptine in Parkinsonism is as :**
    AP 1994
    A. Dopa 1 agonist
    B. Dopa 1 antagonist
    C. Dopa 2 agonist
    D. Dopa 2 antagonist
    (Harrison-2400)

516. **If penicillin it given intrathecally, it causes :** **AP 1995**
    A. Allergy
    B. Nausea and vomiting
    C. Convulsions
    D. Thrombophlebitis

517. **Circular bobbing is caused by lesions in :** **Rajasthan 1989**
    A. Cerebral cortex
    B. Thalamus
    C. Pons
    D. Cerebullum

518. **Most common cause of CVA is :** **AIIMS 1996**
    A. Meningitis
    B. Arterial thrombosis
    C. Venous thrombosis
    D. Embolism **(Harrison-2368)**

519. **The commonest presentation of secondaries in spine is :**
    UPSC 1986
    A. Pain
    B. Retention of urine
    C. Neuropathy
    D. Paraparesis
    (Harrison-2426)

520. **Pulsating tinnitus is seen in :** **PGI 1986**
    A. Glomus jugulare tumor
    B. Glioma of frontal lobe
    C. Tuberculoma of occipital lobe
    D. Lithium toxicity **(Harrison-2444)**

Ans.: 512. D 513. D 514. D 515. C 516. C 517. C 518. B
519. A 520. A

521. Which of the following poisons can cause convulsions ?

AMU 1985

A. Chlorinated hydrocarbons
B. Strychnine
C. Insecticides
D. All of the above

(Harrison-2360)

522. Mad cow disease (eating of infected meat) is incrimated to lead to :

AIIMS 1996; AI 1997

A. Huntington's chorea
B. Creutzfeldt jacob disease
C. Pick's disease
D. Alzheimer's disease

523. Dressing apraxia is due to lesion in : AIIMS 1998

A. Dominant parietal lobe
B. Non-dominant parietal lobe
C. Dominant frontal lobe
D. Non-dominant frontal lobe

(Harrison-143)

524. Increased pain sensation on pressure on the eyes eg-seen in early stages of tabes is known as :   PGI 1985

A. Haenel's sign
B. Pitre's sign
C. Remark's sign
D. Berger's sign

525. Binda's sign (sudden movement of side) is present is :

A. Corticospinal tract lesions   AMU 1986
B. Pyramidal tract lesion
C. Early stages of TB meningitis
D. XII nerve palsy

526. Following signs may be seen in tabes dorsalis except :

AIIMS 1985

A. Abadie's sign
B. Biernacki's sign
C. Remark's sign
D. None of the above

527. The dose of praziquantel in cerebral cysticercosis :   Bihar 1990

A. 50 mg/kg for 10 days
B. 0.5 mg daily for 14 days
C. 50 mg single dose
D. 0.5 mg thrice daily for 21 days

(Harrison-1248)

528. Gait in Parkinsonism is described as :   Kerala 1994

A. Ataxic gait
B. Festinant gait
C. High stepping gait
D. Dancing gait

(Harrison-2399)

529. Anterior branch of left middle cerebral artery blockage in a left handed person causes :   Delhi 1996

A. Nominal aphasia
B. Motor aphasia
C. Seizures
D. Incontinence

(Harrison-2383)

Ans.: 521. D   522. B   523. A   524. A   525. C   526. D   527. A
528. B   529. B

530. Sturge Weber syndrome is characterized by :   Delhi 1996
A. Capillary or cavernous haemangioma
B. Sensorimotor paralysis
C. Occurs always in distribution of trigeminal nerve
D. Focal seizure first neurologic symptom

531. Spastic paraplegia may be seen in all of the following except :
UPSC 1996
A. Lathyrism   B. Motor neurone disease
C. Cervical spondylosis   D. Chronic lead poisoning
(Harrison-2411)

532. A 45-year old male is brought with a history of headache and slowly progressive weakness of the left half of his body of three months duration. Examination reveals normal ocular fundi, dilated non-reacting pupil on the right and aleft hemiparesis. The most likely diagnosis is :   UPSC 1996
A. Right-sided supratentoiral space occupying lesion
B. Syphilitic basal meningitis
C. Multiple sclerosis
D. Aneurysm of right internal carotid artery

533. A patient is having unilateral recurrent attack of headache recurring many times a day and of short duration in association with flushing, sweating of face, rhinorrhoea and lacrimation. The most likely diagnosis is :   UPSC 1996
A. Migraine   B. Histamine cephalgia
C. Trigeminal neuralgia   D. Maxillary sinusitis
(Harrison-2351)

534. A patient had unilateral lower motor neuron type of facial nerve paralysis with loss of taste sensation. The duration of illness has been one month. After taking treatment, he noticed that during food intake, he was having tears from the affected eye. No other neurological deficit was found on examination. The most likely diagnosis is :   UPSC 1996
A. Ramsay-Hunt syndrome   B. Crocodile tears syndrome
C. Millard-Gubler's syndrome   D. Weber's syndrome

535. A young male is seen in Casualty for weakness in all the four limbs with a history of febrile illness of about 10 days duration. On enquiry, he reported that he had difficulty in climbing stairs and combing his hair which increased rapidly in 3 to 4 days time and left him

Ans.: 530. C   531. C   532. A   533. A   534. B   535. B

bedridden. He complained of subjective paraestheisia but showed minimal sensory loss, with more involvement of proximal muscles than the distal ones; there was no bladder and bowel involvement. The plantars were absent. Cremastric reflex and abdominal reflex were present. CSF examination revealed cells 10/cmm, proteins 140 mg%, sugar 60mg%, chlorides 750 mg. The most likely diagnosis is:

UPSC 1996

A. Acute haematomyelia   B. Guillain-Barre syndrome
C. Transverse myelitis   D. Polymyositis

(Harrison-2503)

536. **Gertsman syndrome (seen in lesions of dominant parietal lobes) is characterised by following except :**
A. Alexia   B. Agnosia   **DNB 1995**
C. Agraphia   D. Acalculia

(Harrison-143)

537. **In TB meningitis the number of leucocytes is :**   **B.U. 1996**
A. < 500   B. 500-1000
C. 1000-1500   D. 1500-2000   **(Harrison-1028)**

538. **If nominal aphasia with blindness, lesions are found in —area :**
PGI 1996

A. Temporosphenoidal   B. Parietal
C. Occipital   D. All of the above

539. **In cortical blindness :**   **PGI 1996**
A. Both direct and consensual reflexes are present
B. Absent direct but consensual reflexes are present
C. Absent direct but intact consensual reflex
D. Absent consensual but positive direct reflex

540. **Antalgic gait is :**   **PGI 1996**
A. Stance phase of affected limb decreased
B. Stance phase of affected limb increased
C. Both
D. None of the above

541. **Lathyrism causes :**   **PGI 1996**
A. Pure motor paraplegia   B. Pure sensory paraplegia
C. Both   D. None

542. **Foster Kennedy syndrome involves which lobe :**   **AP 1996**
A. Temporal   B. Occipital
C. Frontal   D. Parietal

Ans.: 536. A   537. A   538. A   539. A   540. A   541. A   542. C

543. Amusia is seen in lesions of ——:   **DNB 1995**
    A. Dominant parietal lobe
    B. Non-dominant parietal lobe
    C. Temporal lobe (dominant)
    D. Non dominant temporal lobe

The following given item consists of two statements, one labelled the 'Assertion A' and the other labelled the 'Reason R'. You are to examine these two statements carefully and decide if the Assertion A and the Reason R are individually true and if so, whether the Reason is a correct explanation of the Assertion. Select your answers to these items using the codes given below and mark your answer sheet accordingly :   **UPSC 1998**
Codes :
    A. Both A and R are true and R is the correct explanation of A
    B. Both A and R are true but R is not a correct explanation of A
    C. A is true but R is false
    D. A is false but R is true   **(Harrison-2459)**

544. Assertion (A) : Projection of vestibulospinal pathway to cerebral cortex via thalamus assists in the maintenance of postural stability.
    Reason (R) : Somatosensory system (from skin, joint and muscles) and visual system (from retina) compensates any deficiency of vestibulospinal pathway.

545. Spasticity in the right leg usually indicates damage to the :   **Orissa 1998**
    A. Right spinocerebellar tract   B. Right spinothalamic tract
    C. Right corticospinal tract   D. Left posterior column
   (Harrison-115)

546. Cerebral oedema in a comatosed patient following carbon monoxide poisoning can be effectively reduced by :   **CSE 1999**
    A. Oral glycerol   B. Corticosteroids
    C. Hypothermia   D. Hyperbaric oxygen

547. Match List I and List II and select the correct answer using the codes given below the lists :   **CSE 1999**

List I (Type of defect)	*List II* (Site of Lesion)
A. Astereognosis	1. Posterior part of in ferior frontal gyrus
B. Global Aphasia	2. Left angular-supra marginal gyrus area

Ans.: 543. C   544. B   545. C   546. D   547. C

C. Motor Aphasia
D. Agraphia

3. Parietal lobe
4. Entire Parasylvian area

Codes :

A. A   B   C   D
    4   3   1   2

B. A   B   C   D
    3   4   1   2

C. A   B   C   D
    3   4   1   2

D. A   B   C   D
    4   3   2   1

548. **Viral meningitis with reduced CSF sugar is seen in :**
                                            Karnataka 1999
A. Rubella
C. AIDS

B. Mumps
D. Cox sackie virus
                                    (Harrison-2482)

549. **The finding in CSF taken from below the level of an intramedullary tumor :**   **Kerala 1999**
A. Low protein
C. High chloride

B. High protein
D. ↑ fibrinogen   **(Harrison-2464)**

550. **Drug useful in both petitmal and grandmal epilepty is :**
                                            Kerala 1999
A. Phenytoin
C. Carbamazepine

B. Na valproate
D. Phenobarb   **(Harrison-2538)**

551. **Saw toothed EEG with superimposed large spikes is seen in:**
                                            Kerala 1999
A. Petitmal
C. Generalised tonic clonic

B. Complex partial
D. Simple partial
                                    (Harrison-2331)

552. **Opsoclonus is :**   **PGI 1999**
A. Regular conjugate eye movements
B. Conjugate chaotic movements
C. Conjugate, chaotic, continuous
D. Chaotic, disconjugate   **(Harrison-639)**

553. **A lesion in the paracentral lobule causes :**   **PGI 1999**
A. Contralateral foot weakness
B. Seizures only
C. Migraine
D. Cognitive loss

Ans.: 548. B   549. B   550. B   551. B   552. A   553. A

554. A medial temporal lesion produces :   **PGI 1999**
- A. Visual amnesia only
- B. Auditory amnesia
- C. Apraxia
- D. Anterograde learning problems

555. Cervical cord injury does not cause :   PGI 1999
- A. Horner's syndrome
- B. Loss of sensation over face
- C. Spasticity of foot
- D. Wasting with fasciculations of lower limb

                           (Harrison-180)

556. Most sensitive part of brain to hypoxia is :   **PGI 1999**
- A. Thalamus   B. Hippocampus
- C. Cerebellum   D. Caudate nucleus

557. Meningitis with normal glucose concentration is :

                           PGI 1999
- A. Tuberculous   B. Listerial
- C. Cryptococcal   D. Coxsackie

                       (Harrison-2481)

558. Not involved in vascular lesion of anterior spinal artery :

                           PGI 1999
- A. Pain and temperature sensation
- B. Sphincters
- C. Touch and pressure
- D. Pyramidal tracts   **(Harrison-2415)**

559. If the pyramid is cut at the medula, which does not occur:

                          AIIMS 1999
- A. Paresis
- B. Incoordination
- C. Inability to take things in the pain
- D. None of the above   **(Harrison-125)**

560. Arun, diagnosed to have epilepsy recently and put on phenytoin. He was previously on antidepressants. He has developed fatigue and anorexia. His Hb=8.0;Tc=7500;ESR=30 in Ist hour; SGOT35; SGPT 35 and Bb 0.6. The best test for diagnosis would be:

                          AIIMS 1999
- A. Chest X-ray   B. Urine culture
- C. MCV   D. Stool examination

Ans.: 554. D   555. A   556. B 557. D 558. C   559. A   560. C

561. **A married female of 39 year age presents with right sided headache for the past 2 year; Associated nausea, vomiting and photosensitivity is present. Which is the diagnosis:**   **AIIMS 1999**
    - A. Cluster headache
    - B. Glaucoma
    - C. Temporal arteritis
    - D. Retinal retachment
    (Harrison-2353)

562. **The commonest aphasia in metabolic encephalopathy:**
    AIIMS 1999
    - A. Anomic
    - B. Transcortical
    - C. Transcortical sensory
    - D. Broca's   **(Harrison-147)**

563. **Patient with fluent speech, but cannot reproduce listened sentenses melodic tone, lesion is present in :**   **AIIMS 1999**
    - A. Dominant, Inf. frontal cortex
    - B. Post temporoparietal cortex dominant hemisphere
    - C. Post temporoparietal non dominant
    - D. Inf. frontal cortex non dominant

564. **As age progresses receptive aphasia progresses with :**
    AIIMS 1999
    - A. PKU
    - B. Down's syndrome
    - C. Alkaptonuria
    - D. Attention deficit disorder

565. **Hypotonia is seen is all except :**   **MAHE 1999**
    - A. LMN lesion
    - B. Cerebellar lesion
    - C. Dorsal Nerve root compression
    - D. Parkinsonism   **(Harrison-2399)**

566. **Findings in central diabetes insipidus are consistent with :**
    AI-2000

	Serum osmolality	*Urine osmilality*
A.	260	50
B.	300	50
C.	50	300
D.	250	500

567. **A 60-year old hypertensive male, with history of several episodes of right monocular blindess (each episode lasting a few minutes) suddenly develops left sided hemiplegia, hemianaesthesia and motor aphasia. Which one of the following arteries is most likely to be involved ?**   **UPSC 2000**
    - A. Middle cerebral artery
    - B. Internal carotid artery
    - C. Posterior cerebral artery
    - D. Anterior cerebral artery
    (Harrison-2372)

Ans.: 561. A   562. A   563. B   564. A   565. D   566. B   567. A

568. Which one of the following does not cause hydrocephalus?
UPSC 2000
A. Tuberculous meningitis
B. Intraventricular tumour
C. Benign intracranial hypertension
D. Subarachnoid haemorrhage (Harrison-124, 2481)

569. Following are seen in cerebellar lesion except : Delhi 2000
A. Pain B. Involuntary movements
C. Dysdiadokinesia D. Ataxia

570. Features of Balint's syndrome include which of the following :
Kerala 2000
A. Prosopagnosia B. Simultanagnosià
C. Visual object agnosia D. Construction apraxia
E. Dressing apraxia (Harrison-139)

571. Choice of calcium channel blocker in subarachnoid haemorrhage is: Karnataka 2000
A. Nifedipine B. Nimodipine
C. Diltiazem D. Verpamil (Harrison-2605)

572. A female has episodic, recurrent headache in left hemicranium with nausea and parasthesia on right upper and lower limbs is most probably suffering from : AIIMS 2000
A. Migraine
B. Brain tumour
C. Herpes zoster infection of trigeminal
D. Glossopharyngeal neuralgia (Harrison-2445)

573. Impotence is a feature of which of the following : AIIMS 2000
A. Multiple sclerosis
B. Poliomyelitis
C. Amyotropic lateral sclerosis
D. Meningitis (Harrison-2452)

574. Extrapyramida! symptoms are not seen in all except : AIIMS 2000
A. Paralysis agitans B. Carbon monoxide poisoning
C. Cerebro vascular accident D. Multiple sclerosis
(Harrison-2399)

575. Which of the following is the most common cause of late neutologicla deterioration in case of cerebrovascular accident : AIIMS 2000
A. Rebleeding B. Vasopasm
C. Embolism D. Hydrocephalus
(Harrison-2388)

Ans.: 568. B 569. A 570. B 571. B 572. B 573. A 574. D
575. B

576. **A 45 years male presents with hypertension. He has abnormal movements in right upper and lower limbs. Mostly likely site of haemorrhage is :** AI 2001
   A. Lateral ventricles
   B. Caudate nuclei
   C. Pons
   D. Subthalamic nuclei
   (Harrison-2386)

577. **Which of the following is not seen in Huntington's disease :**
   A. Cognitive symptoms
   B. Cog wheel rigidity   AI 2001
   C. Chorea
   D. Family history
   (Harrison-2397)

578. **Alzheimer's disease is associated with :** AI 2001
   A. Down syndrome
   B. Marfan's syndrome
   C. Turner's syndrome
   D. Klinefelter's syndrome
   (Harrison-2392)

579. **A patients CSF report is sugar 40 mg%, protein 150 mg%, chloride of 50 meq/L and lymphocytosis present, diagnosis is :** AI 2001
   A. Fungal meningitis
   B. Viral meningitis
   C. TB meningitis
   D. Leukemia meningitis
   (Harrison-1028)

580. **True statement is about neurocysticercosis :** AI 2001
   A. Usually presents with seizure resistant to anti epileptic drugs
   B. Albendazole is more effective than praziquantel
   C. Usually presents with 6th nerve palsy and hemiperesis
   D. Rare in vegetarians   (Harrison-1248)

581. **Calcium channel is defective in :** Jipmer 2001
   A. Hypokalemic periodic paralysis (PP)
   B. Hyperkalemic PP
   C. Thyrotoxic PP
   D. Paramyotonia congenita   (Harrison-2538)

582. **Jargon speech is seen in :** Jipmer 2001
   A. Broca's aphasia
   B. Wernicke's aphasia
   C. Global aphasia
   D. Conduction aphasia

583. **Best drug for hypertensive encephalopathy is :** Kerala 2001
   A. Sod. nitroprusside
   B. Captopril
   C. Nifedipine
   D. Labetolol   (Harrison-1420)

584. **CT scan shows subarachnoid hemorrhage following symptoms of sudden headache & paralysis, next investigation of choice is :**
   A. 4 vessel angiography
   B. SPECT   Kerala 2001
   C. Transacromial doppler USG
   D. MRI

Ans.: 576. D   577. B   578. A   579. C   580. A   581. A   582. A
    583. A   584. A

585. Recurrent seizures can lead to : Kerala 2001
A. Alzheimers disease B. Berry aneurysms
C. Cerebellar atrophy D. None (Harrison-146)
586. Neuroasthiopathy occurs in all except : Rohtak 2001
A. Fredreich's ataxia B. DM
C. Tabes dorsalis D. None (Harrison-2410)
587. Parkinsonian gait : Rohtak 2001
A. Festitant B. Waddling
C. Staggering D. Lurching (Harrison-2399)
588. Drug of choice for meningococcal meningitis is : MAHE 2001
A. Doxycycline B. Penicillin
C. Ceftriaxone D. Erythromycin
(Harrison-2466)
589. A patient with IVDP, his EHL lost, ankle jerk preserved which is
the site of prolapse ? MAHE 2001
A. L4 B. L5/S1
C. L5 D. L3
590. True about prion diseases is all except : AIIMS 2001
A. Prion is an infectious protein
B. Dementia is the presenting feature
C. 10-15% present with myoclonus
D. Brain biopsy is diagnostic (Harrison-2490)
591. For a patient wity parameningeal rhabdomyosarcoma, the best
investigation is : AIIMS 2001
A. CSF cytology B. CECT
C. MRI D. SPECT scan (Harrison-626)
592. A 60-year-old patient, Asim Singh has presented with a history of
dizziness and vertigo with Horner's syndrome of the right side.
There is loss of pain and temperature on the contralateral side. The
artery involved is most likely to be : AIIMS 2001
A. Anterior inferior cerebellar artery
B. Posterior inferior cerebellar artery
C. Middle cerebral artery
D. Spinal artery (Harrison-2376)
593. All are true about Duchenne muscular dystrophy, except :
A. Shoulder girdle is affected first DNB 2001
B. Gower sign positive
C. Muscle biopsy shows increased number of central nuclei
D. Mental impairment may be seen (Harrison-2530)

Ans.: 585. D 586. A 587. A 588. B 589. C 590. C 591. C
592. B 593. C

594. **Dystrophin gene is of great diagnostic value in :**   DNB 2001
  A. Limb girdle dystrophy
  B. Myotonic dystrophy
  C. Duchenne's muscular dystrophy
  D. Oculopharyneal dystrophy   (Harrison-2530)

595. **Neck rigidity without fever is seen in : DNB 2001**
  A. Meningitis   B. Meningism
  C. Subarachnoid hemorrage D. Space occupying lesion

596. **Bleeding in which site causes coma, pinpoint pupil :**   DNB 2001
  A. Pons   B. Thalamus
  C. Medulla   D. Spinal cord   (Harrison-2386)

597. **Muscle hypotonia is seen in :**   Delhi 2001
  A. Tabes dorsalis   B. Chloropromazine toxicity
  C. Pseudobulbar palsy   D. Post encephalitis
                                                 (Harrison-2432)

598. **Crossed aphasia means :**   JIPMER 2002
  A. Right hemispherical lesion in Rt handed person
  B. Right hemipherical lesion in Lt handed person
  C. Left hemispherical lesion in Lt. handed person
  D. Left hemispherical lesion in Rt handed person   (Harrison-140)

599. **A patient presented with ascending symmetrical paralysis with albuminocytological dissociation. Treatment of choice is :**
                                                      AIIMS 2002
  A. Oral prednisolone   B. IV Ig
  C. IV steroids   D. Plasmapheresis
                                                 (Harrison-2507)

600. **All the following are found in brain dead patients except :**
                                                      AI 2002
  A. Decreased DTR
  B. Absent pupillary reflexes
  C. Complete apnea
  D. Heart rate responding to atropine   (Harrison-5)

601. **A 35 year old female patient, Radha, having children aged 5 and 6 years has history of amenorrhea and galactorrhea. Blood examination reveals increased prolactin. The CT of the head is likely to reveal :**   AI 2002
  A. Pituitary adenoma   B. Craniopharyngioma
  C. Sheehan syndrome   D. Pinealoma
                                                 (Harrison-2033)

Ans.: 594. C   595. D   596. A   597. A   598. A   599. B   600. A
   601. A

602. Craniospinal irradiation is useful in which of the following conditions: AI 2002
    A. Pilocytic astrocytoma   B. Oligodendroglioma
    C. Medulloblastoma   D. Oncocytoma   **(Harrison-2446)**

603. Drop metastasis in lumbar CSF are caused by :   **Delhi 2002**
    A. Astrocytoma   B. Medulloblastoma
    C. Ependymoma   D. Meningioma
                            (Harrison-2446)

604. Prion agent is associated with causation of :   **Delhi 2002**
    A. Lymphocytic choriomeningitis
    B. Spongiform encephalopathy
    C. Progressive multifocal leukoencephalopathy
    D. Subacute scleropsing panencephalitis   **(Harrison-2486)**

605. Which of the following not seen in Wernicke's disease :
                                         Delhi 2002
    A. Medial gaze palsy
    B. Global confusion
    C. Ataxia
    D. Conguate and lateral gaze palsy   **(Harrison-2496)**

606. Grasping and hosping is seen in damage to which part of brain :
                                  Maharashtra -2000
    A. Temporal   B. Occipital
    C. Parietal   D. Frontal

607. Fluent speech is feature of damage to :   **Maharashtra -2002**
    A. Wernicke's area   B. Broca's area
    C. Arcuate faciculus   D. None

608. Nucleus acumbence is related to which of the following?
                                  Maharashtra -2002
    A. Basal ganglia   B. Brainstem
    C. Thalamus   D. Cerebellum

609. The commonest site for hypertensive intracerebral bleed is :
                                    AIIMS 2002
    A. Putamen   B. Cerebellum
    C. Pons   D. Midbrain

610. Which of the following is the treatment of choice for cryptococcal meningitis :   **AIIMS 2002**
    A. Fluconazole   B. Itraconazole
    C. Fluocytosine   D. Amphotericin B

Ans.: 602. C   603. B   604. B   605. A   606. D   607. A   608. A
    609. A   610. D

611. Which of the following is not a usual feature of right middle cerebral artery territory infarct :   AIIMS 2002
    A. Aphasia   B. Hemiparesis
    C. Facial weakness   D. Dysarthria

612. The term post-traumatic epilepsy refers in seizures occurring :
    A. Within moments of head injury   AIIMS 2002
    B. Within 7 days of head injury
    C. Several weeks to months after head injury
    D. Many years after head injury

613. Which one of the following is not true regarding Pseudobulbar palsy:
                                                         UPSC 2002
    A. Emotional incontinence   B. Exaggerated jaw jerk
    C. Dysarthria   D. Flaccid tongue

614. The following condtions may present with both upper and lower motor neurone involvement except :   UPSC 2002
    A. Subacute combined degeneration of spinal cord
    B. Friedreich's ataxia
    C. Amyotrophic lateral sclerosis
    D. Becker's Dystrophy

615. Which of the following blood group antigens serves as receptor for Plasmodium vivax :   UPSC 2002
    A. Duffy   B. ABO
    C. Kell   D. Rh

616. Central nervous system manifestations in chronic renal failure are a result of all of the following except :   AI 2003
    A. Hyperosmolarity   B. Hypocalcemia
    C. Acidosis   D. Hyponatremia

617. Which of the following is the commonest location of hypertensive hemorrhage?   AIIMS 2003
    A. Pons   B. Thalamus
    C. Putamen/external capsule   D. Cerebellum

618. Which of the following is the most common central nervous system parasitic infection?   AIIMS 2003
    A. Echinococcosis   B. Sparganosis
    C. Paragonimiasis   D. Neurocysticercosis

619. Which of the following is the most common tumor associated with type I neurofibromatosis?   AIIMS 2003
    A. Optic nerve glioma   B. Meningioma
    C. Acoustic Schwannoma   D. Low grade astrocytoma

Ans.: 611. A   612. C   613. D   614. D   615. A   616. A   617. C
     618. D   619. C

620. **Which of the following is not a usual feature of right middle cerebral artery territory infarct ?**

AIIMS 2003

A. Aphasia   B. Hemiparesis

C. Facial weakness   D. Dysarthria

621. **A 45 years old hypertensive male presented with sudden onset severe headache, vomiting and neck stiffness. On examination he didn't have any focal neurological deficit. His CT scan showed blood in the Sylvain fissure. The probable diagnosis is :**

AIIMS 2003

A. Meningitis   B. Ruptured aneurysm

C. Hypertensive bleed.   D. Stroke

622. **A middle aged woman, on oral contraceptive for many years, developed neurological symptoms such as depression, irritability, nervousness and mental confusion. Her hemoglobin level was 8 g/dL. Biochemical investigations revealed that she was excreting highly elevated concentrations of xanthurenic acid in urine. She also showed high levels of triglycerides and cholesterol in serum. All the above findings are most probably related to vitamin B6 deficiency caused by prolonged oral contraceptive use except?**   AI 2004

A. Increased urinary xanthurenic acid excretion

B. Neurological symptoms by decreased synthesis of biogenic amines

C. Decreased hemoglobin level

D. Increased triglyceride and cholesterol level

623. **A young male develops fever, followed by headache, confusional state, focal seizures and a right hemiparesis. The MRI performed shows bilateral frontotemporal hyperintense lesion. The most likely diagnosis is ?**   AI 2004

A. Acute pyogenic meningitis

B. Herpes Simplex Encephalitis

C. Neurocysticercosis

D. Carcinomatous meningitis

624. **All of the following can cause neuropathies with predominant motor involvement except?**   AI 2004

A. Acute inflammatory demyelinating polyneuropathy

B. Acute intermittent porphyria

C. Lead intoxication

D. Arsenic intoxication

Ans.: 620. A   621. B   622. D   623. A   624. D

625. **The first investigation of choice in a patient with suspected subarachnoid haemorrhage should be?**   **AI 2004**
   A. Non-contrast computed tomography
   B. CSF examination
   C. Magnetic Resonance Imaging (MRI)
   D. Contrast-enhanced computed tomography

626. **Duchenne muscular dystrophy is a disease of ?**   **AI 2004**
   A. Neuromuscular junction
   B. Sarcolemmal proteins
   C. Muscle contractile proteins
   D. Disuse atrophy due to muscle weakness

627. **Consider the following statement with reference to Poliomye-litis :**   **UPSC 2004**
   1. The incubation period of poliomyelitis is 4 to 35 days, but usually it lies between 7 to 14 days.
   2. Tendon reflexes are normal but sensations are not intact.
   3. Paralysis due to heavy metal poisoning can be easily distinguished clinically from poliomyelitis.
   4. The histopathologic findings in poliomyelitis include necrosis of neuronal cells and perivascular cuffing.
   5. The major advantagesof oral polio vaccine ncude ease of adminstration and secondary immunization of nonimmune contacts in the population.
   Which of the above statements are correct ?
   A. 1, 2 and 3   B. 1, 4 and 5
   C. 3, 4 and 5   D. 2, 3 and 4

628. **Which one of the following is not a feature of Sturge-Weber syndrome (Encephalotrigeminal syndrome) ?**   **UPSC 2004**
   A. "Railroad track" calcification
   B. Focal convulsion
   C. Hemiatrophy of brain
   D. Empty Sella

629. **Ropinirole is the most useful for the treatment of :**   **AIIMS 2004**
   A. Parkinson's disease   B. Wilson's disease
   C. Hoffman's syndrome   D. Carpal tunnel syndrome

630. **All of the following statements are true regarding central nervous system infections, except :**   **AIIMS 2004**
   A. Measles virus is the causative agent for subacute sclerosing pan encephalitis (SSPE)

Ans.: 625. A   626. B   627. B   628. D   629. A   630. B

B. Cytomegalo virus causes bilateral temporal lobe hemorrhagic infraction

C. Prions infection causes spongiform encephalopathy

D. JC virus is the causative agent for progressive multifocal leuco-encephalopathy

631. **All of the following drugs are used for managing status epileptics except :**
    A. Phenytoin   B. Diazepam
    C. Thiopentone sodium   D. Carbamazepine

632. **A patient following head injury was admitted in intensive care ward with signs of raised intracranial pressure. He was put on ventilator and started on intravenous fluids and diuretics. Twenty four hours later his urine output was 3.5 liters, serum sodium 1.56 mEq/l and serum osmolarity of 316 mOsm/kg. The most likely diagnosis based on there parameter is :**
    A. High output due to diuretics
    B. Diabetes insipidus
    C. To much infusion of normal saline
    D. Cerebral salt retaining syndrome

633. **A middle aged man present with progressive atrophy and weakness of ands and forearms. On examination he is found to have slight spasticity of the legs, generalized hyper-reflexia and increased signal in the cortico-spinal tracts on T2 weighted MRI. The most likely diagnosis is :**
    A. Multiple sclerosis
    B. Amyotrophic lateral sclerosis
    C. Subacute combined degeneration
    D. Progressive spinal muscular atrophy

634. **Which one of the following is the most common location of hypertensive bleed in the brain ?**   **AI 2005**
    A. Putamen/xternal capcule   B. Pons
    C. Ventricles   D. Lobar white matter

635. **With which one of the following Lower motor neuron lesions are associated ?**   **AI 2005**
    A. Flaccid paralysis   B. Hyperactive stretch reflex
    C. Spasticity   D. Mucular incorrdination

636. **All of the following are neurologic channelopathies except :**
    A. Hypokalemic periodic paralysis   **AI 2005**
    B. Episodic ataxia type 1

Ans.: 631. D   632. B   633. B   634. A   635. A   636. D

    C. Familial hemiplegic migraine

    D. Spinocerebellar ataxia 1

637. **According to the Glasgow Coma Scale (GCS), a verbal score of 1 indicates :**   **AI 2005**
    A. No response   B. Inappropriate words
    C. Incomprehensible sounds   D. Disoriented response

638. **EEG is usually abnormal in all of the following except :**   **AI 2005**
    A. Subacute sclerosing panencephalitis
    B. Locked - in state
    C. Creutzfoldt-Jackob disease
    D. Hepatic encephalopathy

639. **Cluster headache is characterized by all, except :**   **AI 2005**
    A. Affects predominantly females
    B. Unilateral headache
    C. Onset typically in 20-50 years of life
    D. Associated with conjunctival congestion

640. **Which one of the following tumours shows calcification on CT scan:**   **AI 2005**
    A. Ependymoma   B. Medulloblastoma
    C. Meningioma   D. CNS lymphoma

641. **The earliest manifestation of increased intracranial pressure following head injury is :**   **AI 2005**
    A. Ipsilateral pupillary dilatation
    B. Contralateral pupillary dilatation
    C. Altered mental status
    D. Hemiparesis

642. **In Von Hippel-Lindau Syndrome, the retinal vascular tumours are often associated with intracranial hemangioblastoma. Which one of the following regions is associated with such vascular abnormalities in this syndrome?**   **AI 2005**
    A. Optic radiation   B. Optic tract
    C. Cerebellum   D. Pulvinar

Ans.: 637. A   638. B   639. A   640. C   641. C   642. C

RECENT QUESTIONS

1. Self-stimulation is experimentally done from which part of brain :
   A. Medial forebrain bundle
   B. Area arond aqueduct of sylvius
   C. PV region of hypothalamus
   D. Radical radiotherapy

2. Stimulation of which of the following areas of brain is experimentally used to control intractable pain :
   A. Periaqueductal grey matter
   B. Mesencephalon
   C. Subthalamic nucleus
   D. Medial forebrain bundle

3. A politician is shot in the back during a rally at level of T8 vertebra Immediately after the shot he loses all the sensation below level of lesion. Chance of regeneration of spinal cord due to the fact that injured nerve is not able to regenerate is due to following reason except :
   A. Lack of endoneural tubes
   B. Lack of growth·factors
   C. Presence of glial scar
   D. Lack of myelin inhibiting substance

4. Which of the following is true :
   A. Dopamine increases the hepatic and mesenteric blood flow at high dose
   B. Dobutamine decreases peripheral vascular resistance
   C. Nor-adrenaline increases the renal blood flow
   D. Adrenaline causes selective renal vasodilation

5. Which of the following is not true about polymyositis :
   A. Limb girdle weakness
   B. Opthalmoplegia
   C. Para-neoplastic syndrome
   D. Spontaneous discharge in EMG

Ans: 1. A   2. A   3. D   4. B   5. B

6. With ageing, a slight decrease in cognitive impairment is seen due to increase in level of :
   A. Homocysteine
   B. Taurine
   C. Methioine
   D. Cysteine

7. **Blink reflex is used for :**
   A. Mid pontine lesions
   B. Neuromuscular transmission
   C. Axonal neuropathy
   D. Motor neuron disease

8. **Grisel syndrome all are true except :**
   A. Post-adenoidectomy
   B. Conservative treatment
   C. Inflammation of cervical spone ligaments
   D. No need for neurosurgeon

9. All of the following are seen in cervical syringomyelia except :
   A. Burning sensation in hands
   B. Hypertrophy of abductor pollicis brevis
   C. Plantar extensor
   D. Absent biceps reflex

10. Which of the following is a feature of temporal arteritis :
    A. Giant cell arteritis
    B. Granulomatous vasculitis
    C. Necrotizing vasculitis
    D. Leucocytoclastic vsculitis

11. A 70 years old retired military person with good previous medical record complains of bi-temporal headache with scalp tenderness which is decreased in lying down position. He states that he gets relief by giving pressure over bilateral temples. The patient also complains of loss of appetite , weight loss with feeling feverish. Diagnosis is :
    A. Chronic tension headache
    B. Temporal arteritis
    C. Migraine
    D. Fibromyalgia

12. A patient of rehumatoid arthritis develops sudden onset quadriparaesis, Babinsky sign was positive, increased muscle tone of limbs with exaggerated tendon jerks and worsening of gait with no sensory, sphincter involvement. Best initial investigation is :
    A. X-ray of cervical area of neck in flexion and extension
    B. MRI brain
    C. EMG and NCV within 48 hours
    D. Carotid angiography

Ans: 6. A   7. A   8. D   9. B   10. A   11. B
       12. A

13. Gait apraxia is seen in thromboembolic episode involving :
    A. ACA   B. MCA
    C. PCA   D. Posterior choroidal artery

14. A patient with Tubercular meningitis was taking ATT regularly. At end of 1 month of regular intake of drugs deterioration in sensorium is noted in condition of the patient. Which of the following investigations is not required on emergency evaluation :
    A. MRI
    B. NCCT
    C. CSF examination
    D. Liver function tests

15. Which of the following is true about Wilson disease?
    A. High copper in urine, high copper in serum
    B. High ceruloplasmin
    C. Low serum copper
    D. Low urinary copper

16. A neurosurgeon dropped his kid to the school then there he saw a child with uncontrollable lauging and precocious puberty. When he again went to the school in parents teachers meeting, he talked to the father of that boy and advised him to get an MRI done and the diagnosis was confirmed. What is the most probable diagnosis :
    A. Hypothalamic hamartoma
    B. Pineal germinoma
    C. Pituitary adenoma
    D. Craniopharyngioma

17. A child presented to the casualty with seizures. On examination an oval hypo-pigmented macules were noted on the trunk, along with sub-normal IQ. Probabale diangosis of the child is :
    A. Neurofibromatosis
    B. Sturge Weber syndrome
    C. Tuberous sclerosis
    D. Incontinenta pigmenti

18. A preterm infant with poor respiration at birth starts throwing seizures at 10 hours after birth. Anti-epileptic of choice shall be :
    A. Leveteracetam
    B. Phenytoin
    C. Phenobarbitone
    D. Lorazepam

Ans: 13. A   14. C   15. C   16. A   17. C   18. C

19. A 7 years old girl with falling grades and complaints by teacher that she is inattentive in class to her parents and has bad school performance. On hyperventiation her symptoms increased and showed the following EEG findings. Diagnosis is :
    A. Myoclonic epilipsy
    B. Myoclonus
    C. Absence seizure
    D. Juvenile myoclonic epilepsy

20. In a child, CSF examination is not used in diagnosis of :
    A. ALL
    B. Hodgkin's lymphoma
    C. Non-Hodgkin's lymphoma
    D. AML

21. The nucleus involved in Papez circuit is :
    A. Pulvinar
    B. Intralaminar
    C. VPL nucleus
    D. Anteiror nucleus of Thalamus

22. Opioid tolerance develops to all of the following actions, except :
    A. Miosis
    B. Analgesia
    C. Euporia
    D. Nausea and Vomiting

23. All of the following statements about Epidural opioids are true, except :
    A. Acts on the dorsal horn substantia gelatinosa
    B. May cause pruritis
    C. May cause Respiratory Depression
    D. Gastrointestinal adverse effects are not seen

24. Which of the following drug is most commonly used world wide in maintenance doses for opioid dependence :
    A. Naltrexane
    B. Methadone
    C. Imipramine
    D. Disulfiram

25. A young girl presents with repeated episodes of throbbing occipital headache associated with ataxia and vertigo. The family history is positive for similar headaches in her mother. Most likely diagnosis is :
    A. Vestibular Neuronitis
    B. Basilar migraine
    C. Cluster headache
    D. Tension headache

Ans: 19. C   20. B   21. D   22. A   23. D   24. B
      25. B

26. All of the following drugs are used in prophylaxis of migraine, except :
    A. Propanolol   B. Flunarizine
    C. Tapiramate   D. Levetiracetam

27. All of the following are true about anterior choroidal artery syndrome except :
    A. Hemiparesis
    B. Hemisensory loss
    C. Homonymous Hemianopia
    D. Involvement of anterior limb of internal capsule

28. Which of the following sites is not involved in a posterior cerebral artery infarct :
    A. Midbrain   B. Thalamus
    C. Temporal lobe   D. Anterior Cortex

29. All of the followng are true about delirium tremens, except :
    A. Visual Hallucinations   B. Tremors
    C. Ophthalmoplegia   D. Clouding of consciousness

30. A 60 year old man with progressive dementia of recent onset presents with intermittent irregular jerky movements. EEG shows periodic sharp biphasic waves. The most likely diagnosis is :
    A. Alzheimer's disease
    B. Creutzfeldt Jakob disease
    C. Lewy body dementia
    D. Herpes Simplex Encephalitis

31. A thirty five year old female has proximal weakness of muscles, ptosis and easy fatigability. The most sensitive test to suggest the diagnosis is :
    A. Muscle Biopsy   B. CPK levels
    C. Edrophonium test   D. Single fiber EMG

32. NARP syndrome is a type of :
    A. Mitochondrial function disorder
    B. Glycogen storage disorder
    C. Lysosomal storage disorder
    D. Lipid storage disorder

33. Lesions of the lateral cerebellum cause all of the following except :
    A. Incoordination
    B. Intention tremor
    C. Resting tremor
    D. Ataxia

Ans: 26. D   27. D   28. D   29. C   30. B   31. D
       32. A   33. C

34. Low CSF protein may be seen in all of the following conditons, except :
    A. Recurrent Lumbar Puncture
    B. Hypothyroidism
    C. Pseudotumor cerebri
    D. Infants

35. Which of the following drugs should not be used with Rivastigmine in patients with Alzheimer's disease :
    A. SSRI
    B. Tricyclic Antidepressant
    C. RIMA
    D. Atypical Antidepressants

36. All of the following statements about Phenytoin are true, except :
    A. Follows saturation kinetics
    B. Is teratogenic
    C. Is highly protein bound
    D. Stimulates Insulin secretion

37. Serotonin syndrome may be precipitated by all of the following medications, except :
    A. Chlorpromazine
    B. Pentazocine
    C. Buspirone
    D. Meperidine

38. A young male presents with meningiococcal meningitis and allergy to penicillin. Which is the mot suitable drug :
    A. Chloramphenicol
    B. Meropenem
    C. Ciprofloxacin
    D. Teicoplanin

39. Plasmapharesis is used in all of the following except :
    A. Myaesthenic crisis
    B. Cholinergic crisis
    C. Gullian Barre syndrome
    D. Polymyositis

40. A patient presents with ataxia, urinary incontinence and dementia. The likey diagnosis is :
    A. Alzheimer's Disease
    B. Parkinson's disease
    C. Steel Richardson syndrome
    D. Normal Pressure Hydrocephalus

41. Which of the following is the classical CSF finding seen in TBM :
    A. Increased protein, decreased sugar, increased lymphocytes
    B. Increased protein, sugar and lymphocytes
    C. Decreased protein, increased sugar and lymphocytes
    D. Increased sugar, protein and neutrophils

Ans: 34 B   35. B   36. D   37. A   38. A   39. B
      40. D   41. A

42. A 25 years old person presents with acute onset of fever and focal seizures. MRI scan shows hyperintensity in the temporal lobe and frontal lobe with enhancement. The most likely diagnosis is :
    A. Meningococcal Meningitis
    B. Herpes simplex Encephalitis
    C. Japanese Encephalitis
    D. Pick's disease

43. Which of the following represents the site of lesion in Motor Nurone disease :
    A. Anterior Horn cells   B. Peripheral Nerve
    C. Spinothalamic Tract   D. Spinocerebral tract

44. All of the following are true about Guillain Barre Syndrome (GBS), except :
    A. Ascending paralysis
    B. Flaccid paralysis
    C. Sensory involvement
    D. Albumino-Cytological Dissociation

45. Which of the following statements about vasomotor centre (VMC) is true :
    A. Independent of corticohypothalamic inputs
    B. Influenced by baroceptor signals but not by chemoreceptors
    C. Acts along with the cardiovagal centre (CVC) to maintain Blood pressure
    D. Essentially Silent in sleep

46. Action Potential is initiated at the Axon Hillock-Initial segment of the neuron because :
    A. Threshold for Excitation is lowest
    B. Neurotransmitter is released at this site
    C. It is an unmyelinated segment
    D. Has lowest concentration of voltage gated sodium channels

47. Appetite is stimualted by all of the following peptides, except :
    A. Agouti-Related Peptide (AGRP)
    B. Melanocye Stimulating hormone (MSH)
    C. Melanin Concentrative Hormone (MCH)
    D. Neuropeptide Y

48. Key Regulators of sleep are located in :
    A. Hypothalamus   B. Thalamus
    C. Putamen   D. Limbic cortex

Ans: 42. B   43. A   44. C   45. C   46. A   47. B
      48. A

49. Adverse effects of valproic acid derivatives include the following except :

A. Alopecia   B. Liver Failure

C. Weight gain   D. Osteomalacia

50. An elderly male presents with headache, recurrent infections and multiple punched out lytic lesions of X-ray skull. The investigation that will best help in establishing a diagnosis is :

A. Protein Electrophoresis

B. Serum calcium

C. Alkaline phosphatase levels

D. Acid phosphatase levels

51. Restless leg syndrome (RLS) is seen in :

A. Hypercalcemia   B. Hyperphosphatemia

C. Chronic renal failure   D. Hyperkalemia

52. All of the following statements about temporal arteritis are true, except :

A. More common in females

B. Worsens on exposure to heat

C. Seen in elderly women

D. Can lead to sudden bilateral blindness

53. Spastic paraplegia is caused by all, except :

A. Vitamin B_{12} deficiency

B. Cervical spondylosis

C. Lead poisoning

D. Motor neuron disease

54. Which of the following statements about Lambert Eaten Myaesthenic syndrome is true :

A. Tensilon test is positive

B. Extraocular muscles are most commonly involved

C. Incremental response to repeated electrical stimulation

D. Associated with adenocarcinoma of lung

55. Triad os Tuberous Sclerosis includes all. except :

A. Epilepsy   B. Adenoma sebacium

C. Low intelligence   D. Hydrocephalus

56. What is the effect of moderate exercise on cerebral blood flow:

A. Does not change   B. Increases

C. Decreaes   D. Initially inc.eaes and then decreases

Ans: 49. D   50. A   51. C   52. B   53. C   54. C

       55. D   56. A

57. Which of the following sensation are transmitted by the Dorsal Tract/ Posterior column:

A. Fine touch   B. Pain

C. Temperature   D. All of the above

58. Use of Ergotamine is contraindicated in :

A. Diabetes mellitus

B. Anemia

C. Ischaemic heart disease

D. PPH (Postpartum haemorrhage)

59. Flumazenil is a :

A. Diazepam antagonist

B. Diazepam Reverse agonist

C. Agonist at BZD receptors

D. Selective Serotonin Reuptake Inhibitor (SSRI)

60. Dying back neuropathy is seen in all except :

A. Diabetic Neutropathy

B. Arsenic neuropathy

C. Porphyria

D. Gullian Bare syndrome (GBS)

61. Which of the following agent is least likely to cause meningitis in the elderly :

A. Listeria Monocytogenes

B. Streptococcus pneumonia

C. Gram Negative bacteria

D. Herpes Simplex Virus - 2 (HSV 2)

62. Pick's body in Pick's disease is :

A. Tau protein   B. Alpha synuclein

C. Beta synuclein   D. A β amyloid

63. Biondi Ring Tangles (BRT) are found in :

A. Choroidal plexus cells

B. Golgi type II cells

C. Basket cells

D. Piamatter

64. Which of the following is not a feature of extramedullary tumour :

A. Early Corticospinal signs and paralysis

B. Root pain or midine back - pain

C. Abnormal CSF

D. Sacral sparing

Ans:	57. A	58. C	59. A	60. D	61. D	62. A
	63. A	64. D				

65. Clinical features of conus medullaris syndrome include all of the following except :
   A. Plantar Extensor
   B. Absent knee & ankle jerks
   C. Sacral anesthesia
   D. Lower sacral & coccygeal involvement

66. All of the following statements about Diffuse Axonal Injury (DAI) are true except :
   A. Caused by shearing force
   B. Predominant white matter haemorrhages, in basal ganglion and corpus callosum
   C. Increased Intracranial tension is seen in all cases
   D. Most common at junction of grey and white matter

67. Which of the following is not seen in Chronic Regional Pain Syndrome :
   A. Anhydrosis   B.   Pain
   C. Swelling   D.   Osteoporosis

68. Broca's area is concerned with :
   A. Word formation   B.   Comprehension
   C. Repetition   D.   Reading

69. The mechanism of hearing and memory, include all except :
   A. Changes in level of neurotrnasmitter at synapse
   B. Increasing protein synthesis
   C. Recruitment by multiplication of neurons
   D. Spatial Reorganization of synapse

70. The processing of short term memory to long term memory is done in :
   A. Prefrontal cortex   B.   Hippocampus
   C. Neocortex   D.   Amygdala

71. Which of the following is not an antiepileptic agent :
   A. Phenytoin   B.   Topiramate
   C. Flunarazine   D.   Carbamazepine

72. All of the following antiepileptic agents acts via Na+ channel except :
   A. Vigabatrin   B.   Phenytoin
   C. Valproate   D.   Lamotrigne

73. SIADH is associated with the following drug :
   A. Vincristine   B.   Erythromycin
   C. 5-FU   D.   Methotrexate

Ans: 65. B   66. C   67. A   68. A   69. C   70. B
      71. C   72. A   73. A

74. Increased ICT is associated with all except :
    A. Paraparesis
    B. Abducent paralysis
    C. Headache
    D. Visual blurring

75. Pontine Stroke is associated with all except :
    A. Bilateral pin point pupil
    B. Pyrexia
    C. Vagal palsy
    D. Quadriparesis

76. Millard Gubler syndrome includes the following except :
    A. 5th nerve palsy
    B. 6th nerve palsy
    C. 7th nerve palsy
    D. Contralateral hemiparesis

77. All of the following statements are true about Benedict's syndrome, except:
    A. Contralateral tremor
    B. 3rd nerve palsy
    C. Involvement of the penetrating branch of the basilar artery
    D. Lesion at the level of the pons

78. The following are components of Brown Sequard syndrome except :
    A. Ipsilateral extensor plantar response
    B. Ipsilateral pyramidal tract involvement
    C. Contalateral spinothalamic tract involvement
    D. Contralateral posterior column involvement

Ans: 74. A   75. C   76. A   77. D   78. D

REFERENCES
(For Further Reading)

1. Merrit's Textbook of Neurology, (Ed.) Rowland L.P. Lea and Febiger, Philadelphia.

2. Brain's Diseases of the Nervous system Brain L, Walton JN, Oxford University Press, London.

3. Harrison's Principles of Internal Medicine (Eds) Wilson JD, Braunnwald E, Isselbacher KJ, Petersdorf RG, Martin JB, Fauci AS, Root RK. McGraw Hill, Inc.

4. Davidson's Principle and Practice of Medicine. (Ed.) Macleod, J. Churchill Livingstone.

5. API Textbook of Medicine. (Eds) Datey KK, Shah SJ. Association of Physicians of India, Bombay.

6. Rypin's Medical Licensure Examination, (Ed). Frohlich ED. J.B. Lippincott Co., Philadelphia.

7. A Textbook of Radiology and Imaging (Ed.) Sutton D. Churchill Livingstone, Edinburg.

8. Mackie and McCartney Medical Microbiology. (Eds.) Duguid JP, Marmion BP, Swain, RHA. Churchill Livingstone, Edinburg.

9. Pathologic Basis of Disease. Robbins SL, Cotran RS, Kumar V. W.B. Sauders Company, Philadelphia.

10. Grant's Method of Anatomy. (Eds) Basmajian JV, Slinecker CE. Williams and Wilkins, Baltimore.

11. Bhatia's Dictionary of Psychiatry, Psychology and Neurology. (Ed) Bhatia M/s CBS Publishers and Distributors, Delhi, India.

12. Delhi MD/MS Entrance Examination Model Test Papers. (Ed) Bhatia, M/s CBS Publishers and Distributors, Delhi, India.